Computer Applications to Private Office Practice

Computer Applications to Private Office Practice

Edited by

Byron B. Oberst, M.D., F.A.A.P.
Robert A. Reid, M.D., F.A.C.P.

With Contributions by
Elmer Gabrieli, John H. Hoskins, John M. Long,
Gretchen Murphy, Byron B. Oberst, Robert A. Reid

With a Foreword by William A. Bauman, M.D.

Springer-Verlag
New York Berlin Heidelberg Tokyo

Byron B. Oberst, M.D., F.A.A.P.
Omaha Children's Clinic, P.C.
Omaha, Nebraska 68144
U.S.A.

Robert A. Reid, M.D., F.A.C.P.
Department of Internal Medicine
University of Virginia Medical Center
Charlottesville, Virginia 22901
U.S.A.

Library of Congress Cataloging in Publication Data

Main entry under title:
Computer applications to private office practice.

Bibliography: p.
Includes index.
1. Medicine—Data processing. 2. Medical offices—
Data processing. 3. Medicine—Practice. I. Oberst,
Byron B. II. Reid, Robert A.
R858.C628 1984   651.5'04261   83-20270

With 1 Figure

© 1984 by Springer-Verlag New York Inc.
Softcover reprint of the hardcover 1st edition 1984

Typeset by Ampersand, Inc., Rutland, Vermont

9 8 7 6 5 4 3 2 1

ISBN-13: 978-1-4612-9746-8   e-ISBN-13: 978-1-4612-5226-9
DOI: 10.1007/978-1-4612-5226-9

To all those confused, bewildered, and betwixt physicians in solo and small group practices who are indecisive, uncertain, and overwhelmed by the vicissitudes of practice and are considering turning to the computer for solace and rescue

Foreword

This publication is sponsored by the American Association for Medical Systems and Informatics. The Board of AAMSI and the Board of the Society for Computer Medicine, one of AAMSI's predecessors, agreed that a book on application of medical systems and informatics for the practitioner would help promote high quality health care and they charged the Committee on Standards of the Society for Computer Medicine to write such a text. It is intended as a guide for the field of medical systems and informatics with emphasis on standards, terminology, and coding systems.

The text, a result of three years of research and effort, has been reviewed by the Board of Directors of AAMSI and approved by the Publications Committee. We believe that you will find it valuable and hope to revise it from time to time to meet current needs.

On behalf of the members of the Association, we congratulate the authors and thank them for their efforts.

WILLIAM A. BAUMAN, M.D.
President
American Association for
Medical Systems and Informatics

Preface

This book has been written by the members of the Committee on Standards of the Society for Computer Medicine. We have drawn upon the Society's expertise to prepare an easy-to-read and understandable How-to-Do-It text for use by those physicians who are considering computerization of their office in one manner or another.

More physicians are seeking the where, why, what, and how of computer applications to both the business side and the clinical/patient care side of practice management. The general theme of this book is intended for those individuals in solo or small group practices such as two to eight physicians who want help in starting their search for "greener pastures."

Basic principles for the initial approach to office computer applications are defined. Major elements and areas for consideration are outlined. In each chapter the specific author's editorial feelings are included. Some material may be reemphasized elsewhere in the text for impact of importance.

This book *will not* make the reader a computer applications expert or provide a complete do-it-yourself method. This book *will* provide ideas and concepts for consideration, and highlight pitfalls to be avoided.

The committee has greatly enjoyed planning, discussing, and writing this book. We hope that it will help you avoid headaches, heartaches, dissatisfaction, and costly mistakes. We have attempted to discuss our view of the near future of office computers without being carried away by "blue sky." All of us are pragmatists and realists. We welcome you to contact any of the authors or Society members to obtain more information. The American Association for Medical Systems and Informatics welcomes you to the future of health care delivery, office management, and office viability.

Contents

Contributors

Elmer Gabrieli, M.D., PRAKTICE, Computers Serving Medicine, Buffalo, New York

John H. Hoskins, Editor and Publisher, The Medical Practice Letter, New Haven, Connecticut

John M. Long, Ed.D., Director, Coordinating Center, Hyperlipidemia Program, University of Minnesota, Minneapolis, Minnesota

Gretchen Murphy, R.R.A., Assistant Professor, Health Information Services, Seattle University, Seattle, Washington

Byron B. Oberst, M.D., F.A.A.P., Omaha Children's Clinic, P.C., Omaha, Nebraska

Robert A. Reid, M.D., F.A.C.P., Associate Professor, Department of Internal Medicine, University of Virginia Medical Center, Charlottesville, Virginia

Contents

Part I

Introduction
to Accounting Systems

Chapter 1

Beyond Billing: Some Things You Should Know Before You Begin

John M. Long

Objectives

Provide the private practice physician with a basic perspective needed to intelligently evaluate automation for the office, and warn him of some pitfalls.

Provide the physician with some general information he will need to automate parts of his private practice office system, including basic facts about equipment needs, personnel training, service needs, and related items.

Introduction

The medical community has been slow to accept and use automation. For many years automation has been confined primarily to hospital business office applications, such as billing, scheduling, inventory, bed census, and the like. Its impact on private practice has been almost negligible, and primarily confined to billing and a few office management aids. The potential for automation in the clinical practice of medicine has yet to be achieved. The reasons are understandable:

Medical practice is of necessity conservative. Changes come about only after they have been subjected to careful legal, moral, ethical, and scientific evaluation, and are proven beneficial; at least that is how it works in the ideal situation.

Medical practioners are busy, self-directed people who have little time for new and unproven concepts unless a new and unproven concept has a personal appeal.

There has been a serious communication gap between the computer specialist and the medical practitioner. Computer specialists have had great difficulty understanding how the practice of medicine really works, and therefore in understanding what the needs for automation, and its limitations, really are. The problem has been aggravated by a tendency to gloss over many of the real complexities related to automation.

Finally, and perhaps most important, the medical profession is steeped in the personalized one-to-one physician–patient relationship. Many physicians see automation as a threat to this relationship, and well it may be, unless the profession makes sure that it isn't.

The potential for automation in medical practice is nevertheless being recognized by more and more physicians. The 1980s will see a great deal more use of automation in medical *practice*. The word practice is stressed to emphasize clinical uses beyond billing.

There is not yet any real pressure on the physician in private practice to automate his office in order to keep up. This probably will change over the next 10 years. Here are some of the reasons why the climate is becoming more favorable to private practice office automation.

The tremendous increase in paperwork required by governments and insurance carriers has pressured many offices into some automated method, at least for billing. Once this is done, the potential for other uses becomes more obvious, and is easier to accomplish.

Many doctors in practice today have seen the beneficial uses of automation in their training programs; the countless experimental and operational applications which are running in medical schools.

Similarly, many hospitals, whose original uses of computers were business oriented, are expanding into clinical uses.

Many computer and computer service vendors, seeing the tremendous business potential in medical practice, are promoting their uses.

If you noticed that the word computer was used for the first time in the preceding sentence, you should know that this is deliberate, because at this stage a broader perspective is needed. You should be thinking about the process, *automation*, and not the means to accomplish the process, *computers or a computer service*. As an analogy, at this point we want to talk about transportation, not cars.

In this chapter we consider some motivations a physician might have to begin automating some parts of the office, what computers can and cannot do for you, and some important caveats. In the final sections, we discuss some of the things you need to know about your own practice. There may be a few surprises in these latter remarks.

What Automation Can Do for You Today

Definition: In the context of this book, *automation* means the use of some modern technologies (primarily computers, computer services, communication networks, and terminals of various kinds) to assist the physician in the operation of his private practice office. Billing and practice management are certainly the primary uses today. They may well be the only cost-effective uses at this time. Without misleading you, or promising more than can be delivered, we hope to expand your thinking beyond these mundane, but extremely important applications.

It is not unrealistic to consider automating at least key parts of the medical record system. Once the medical record is automated, its use for managing patient education programs, or for monitoring follow-through on a long course of treatment, is possible. The synopsis of a patient's medical record can be produced. Consulting reports can be partially automated. Computers can be used for taking histories, for quality assessments, and for many more things, limited only by the contents of your database, programming requirements, and your imagination. The system can help with the physician's professional development, practice planning, and similar activities. It is even possible, since it has been done, to have a so-called paperless physician's office!

Does all this sound easy? Don't believe it! It doesn't come cheap, either. It all can be done today if you are willing to devote time, energy, and money to do it, but it may or may not be cost-effective. Therefore, you should stop to examine your motives for entering into such an endeavor.

What Are Your Motives?

Legitimate motives include such business reasons as reducing overhead costs or cutting down on paperwork. They include professional reasons such as improving patient management. They can also include very personal reasons as well. Depending on your circumstances, the business reasons may be compelling. The professional reasons are not compelling today, but certainly will be at some time in the future. Taking each set in turn:

Business Reasons

(1) Improve office organization and systems
(2) Reduce the time devoted to the essential but repetitive clerical functions performed by the office staff; especially to reduce the time the doctor devotes to such tasks
(3) Improve the scheduling system
(4) Reduce overhead costs
(5) Improve billing and collections
(6) Reduce lost bills

Professional Reasons

(1) Improve patient management, i.e., management of the follow-through on a course of treatment
(2) Keep track of patient participation in well-care systems
(3) Produce consulting reports in a more timely fashion
(4) Have rapid and easy access to medical records needed by the doctor, the staff, or the patient
(5) Design a planned new office so that it will be functional in the future

Personal Reasons

(1) Desire to be up to date
(2) Personal interest in computers and/or automated systems; for example, you feel such a project would be fun, and a challenge
(3) Desire to influence the direction of these developments in your profession
(4) Enjoy being a pioneer
(5) Desire to give prestige to your practice

There is nothing wrong with any of the personal reasons as long as you recognize them for what they are. Actually, unless one of these personal (or other similar) reasons apply to you, you may not be ready to automate your practice, with the possible exception of basic billing and practice management functions. Such projects today require dedication and commitment of time, energy, and money. The financial payout is uncertain. The rewards may come primarily from the satisfaction of personal motives.

Some Caveats

A computer *can* do almost anything, just as any child born in the United States can be its president. Both statements are true but not very realistic.

Here are some of the things automated systems *cannot* do in the real world of the typical private practice.

(1) You cannot automate a nonsystem. Computers are not human! They run on complex *logic* and require rigid adherence to that logic. If the system you wish to automate (billing, scheduling, records, etc.) is disorganized, it must first be organized before it can be computerized. It often happens that, once the system is well organized and running manually, the need to automate it disappears.

A somewhat haphazard manual system can be made to work by humans, but not on a computer. Humans are flexible, and can make a less-than-perfect system work. Computers have only the flexibility that is programmed into their logical systems. Small automated office systems probably cannot afford much flexibility; thus, the systems need to be well organized.

(2) Automated systems can handle only those situations that they are programmed to handle. All exceptions to a system much be anticipated in an automated system. This can be done by providing manual overrides so that a human can take care of the exceptions not provided for in the system, but *these* must be planned as well.

(3) It is not possible to evaluate a system properly by reading vendor literature, or by viewing a demonstration. This does not imply any dishonesty. It is true, however, that demonstrations are usually carefully orchestrated and that the literature is optimistic. The core problem is one of communication. It relates to both the vendor not really understanding what you want and you not really understanding what he is saying.

(4) It is important to get past the "gee-whiz" phase with computers. We hear that computers send rockets into outer space and decode the complex signals which are returned to create pictures of what they "see." These billion dollar systems are amazing and demonstrate the tremendous potential of computers. The relevant issue for you is to understand the real-world capabilities of a computer or a computer service you can afford!

Essential Elements of a Physician's Office System

Every physician's office has a number of *systems* which operate more or less simultaneously. Usually, each system consists of a set of policies and procedures defining how the system operates, a means for providing a service or function (such as space and equipment), forms or some other means of recording data, files or some other means for storing data, a means of finding and retrieving data, and, most important, personnel trained to make the system operate properly. The major components of a well-defined system are listed in Table 1-1.

For example, the appointment system in a small practice might consist of a book where appointments are recorded, a phone, and a receptionist who

Table 1-1   The Anatomy of a System

---

 (1) Policies and procedures
 (2) Description of flow and volume of patients through system (if relevant)
 (3) Description of flow and volume of material (specimens) through system (if relevant)
 (4) Description of flow and volume of data through system
 (5) Means of entry into system
 (6) Space and equipment needs (include maintenance)
 (7) Staff requirement (level of training required and retraining needs)
 (8) Provisions of service—Protocols
 (9) Recording relevant data
(10) Storage and retrieval of data
(11) Means of tracking both costs and charges
(12) Interaction with other office systems

---

takes calls and makes appointments. She follows a set of rules, written or spoken, provided her by each doctor. The rules might be: Dr. A is off on Tuesdays, Dr. B on Wednesdays; no appointments before 10 A.M. to allow for hospital rounds; Dr. C goes to Rotary every Friday from 12 to 1:45 P.M., and so on.

When all of the interlocking systems in the office have been described, you have defined the office. The major systems with potential for automation in a typical small multiphysician office are discussed in subsequent chapters.

The list in Table 1-1 is not intended to be comprehensive or even entirely accurate. It is designed to emphasize the comprehensive nature of typical office systems. In manual systems, operated by humans, these components are often taken for granted, but in automated systems they *must* be defined.

Almost without exception, when you automate one of your office systems, you will spend up to 80–90% of your time clarifying the components of the system. It is often also true that 80–90% of the payoff of such an endeavor will come from this exercise, rather than from the automation itself!

The point is that a manual system must be clearly understood before you attempt to automate it, and before you can intelligently buy an automated version from someone else. Often this exercise will require the imposition of rules and discipline for the manual system which may obviate the need to automate. Note that many of the items in the list are not relevant to automation (1–4 and 6–8). Even those items susceptible to automation (5 and 9–11) must be clearly understood in the manual version before they can be automated. Once a manual system is clearly defined, however, automation, if desired or needed, is *relatively* easy to do.

Thus, when you talk of "going onto a computer" or "automating" certain office systems, what you really mean is that you will use a *computer* (now we can talk about the means) to automate certain parts of that total system. Computers are very good at doing highly repetitive tasks and are useful in collecting, storing, manipulating, and retrieving data. In this context, *data* is used in a broad sense to mean information about patients, procedures, providers, and insurance, as well as rules for manipulating the data, descriptions of reports, and when to produce them. As we have implied, you need to know far less about computers than you need to know about your office systems. There are certain basic concepts that need to be understood, nonetheless. The next section provides a brief discussion about the most important of them.

Computer Concepts

The physician does not really need to know very much about a computer in order to use it intelligently. How many people who drive a car understand the internal combustion engine, the electrical system, the braking system, or the steering mechanism? Drivers do need to know how to start a car, steer it, and know when it is operating properly. They need to know the rules of the road, and certain other information is desirable. For example, it is good to know about defensive driving.

The first thing to know about a computer is that is consists of *hardware*, that is, boxes containing electronic circuits, cables, and the like, and *software*, that is, sets of logical statements or *programs* which cause the electronic circuits to modify their states in such a way that logical rules are followed. The hardware is modular, with each component performing a certain function analogous to the component systems (e.g., the engine, electrical, braking, and steering systems) of a car.

The software has been expressed in *languages*, which allow us to communicate with the computer. Machine language, the most basic level of communication, is very tedious, consisting essentially of a string of binary numbers (that is, zeros and ones). Over the years computer languages have become more sophisticated, and most programs are expressed in the so-called higher-level languages. FORTRAN is a language similar to algebra. COBOL is more like English, with terms useful for commercial data processing. You will hear about BASIC and MUMPS. If you don't plan to do your own programming, the programming language used in a system is a matter of relative indifference. People who do program have preferences of theological intensity.

It is important to recognize that there is absolutely no magic in the way computers work. Every single thing they do follows in excruciating detail a preplanned logical sequence. Fortunately, such details are becoming more and more transparent to an end user such as yourself. Many of today's systems have languages and/or programs which allow people with virtually no knowledge of computers to communicate using English-like expressions. The use of numerical codes and other non-English-like elements is still quite common nonetheless.

Appendix A contains material which will provide you with a general understanding of the hardware and software of a computer, and introduce you to some of the jargon. You can skip it entirely if you wish, without any lessening of your ability to understand the remaining discussions in the book, or to proceed with your office automation project. The book will be useful to those of you who are thinking of purchasing one of the so-called personal computers, if your consultant cannot talk without using the jargon, or if a vendor confuses you (sometimes deliberately) with the terminology.

Concluding Remarks

The material presented here has made the following critical points:

(1) The private practice office has many systems which are vital components or modules of the overall office system. Certain parts of these can be automated using a computer, but only after the system is clearly defined and operating in a manual mode.
(2) It is sometimes true that once a system is clearly defined and operating in a manual mode the felt *need* for automation goes away.
(3) One's motivation to automate needs to be clearly understood from the beginning, and the rewards defined in those terms. If your motives are based solely on sound business reasons, i.e., cost savings, you may find little if any payoff for some of the components discussed in this book. If your motives are professional or personal, automation today will be more attractive to you.
(4) Eventually, many functions of the doctor's office will be automated, including the functions described in subsequent chapters of this book. There are no compelling reasons to automate them today, except possibly for billing and related functions. I believe the tide will change within 10 years.
(5) You don't need to be a computer expert in order to effectively use a computer, even if you wish to pioneer its use for certain office practice systems.

The material in this chapter is certainly not designed to destroy interest in the automation of the private practice office. Indeed, I believe the trend toward automation is inevitable! We do hope that it will encourage participation in the trend, but from an informed and realistic perspective.

Chapter **2**

What Sort of Data Should Be Collected and Why?

Robert A. Reid

Objectives

Help the physician think through the long-range goals of an office system. Examine planning for financial amortization and causes of system obsolescence.

Help the physician evaluate the potential of a specific system to meet set goals through examination of the data elements which it handles.

Introduction

During the 1970s most physicians who installed computers in their offices were interested initially in streamlining their accounts receivable. Once billing was operating smoothly, most hoped (at least vaguely) to subsequently use their computers for patient care activities. In general, these hopes for clinical utilization of office computers have not been realized. The major reason for this failure is that the system had unexpected limitations to its functionality. These limitations could have been discerned ahead of time if adequate attention had been given to the data elements which were recorded in the business system itself, and to those data elements which could be recorded at a later date.

The purpose of this chapter is to help you think through your goals for implementing a computer. You should identify the data elements that will be necessary to support these goals. Once these goals are clear, you will, in all probability, initially install an accounts receivable system. Knowing the data elements which you require will help you to examine alternative systems in terms of their potential for expansion.

How Long Will Your System Last?

When establishing goals, it is important to project a time frame for those goals. When installing a computer should you plan for 1 year, for 5 years, or more? Unlike most office equipment, the computer which you buy in the 1980s will have a finite useful life. Your EKG machine or spirometer may still be useful in a decade, but 10 years from now your office computer will be ready for the junk heap.

How will obsolescence occur? Aren't these machines getting more reliable every day? Won't they hold up fine? Of course. Your machine itself will probably function just fine in 10 years—but you will still junk it because it will be obsolete.

The first cause of computer obsolescence is that the power of computers per dollar doubles or triples every year. This does not mean that you can obtain a computer every year which is twice as powerful as the last. Computer power on the commercial market increases in sudden increments. Five years from now you'll be able to buy a machine 30–130 times more powerful than the machine you buy today—for the same money. Does your $10,000 machine handle your accounts receivable now? In 7 years your colleague will spend $10,000 and get a machine which will handle his accounts receivable, his word processing, his investment portfolio, teach his children, and simultaneously support a terminal in every examination room. You will have a Model T, purchased for the same money that he has used to buy a Porsche.

The second cause of obsolescence is declining maintainability. In general computer manufacturers charge 10% of the cost of a machine to keep the machine repaired. Five years from now machines will be designed to be more reliable. They will have redundant parts which automatically take over when there is a malfunction. They will have fewer and fewer parts as more circuitry is placed on a single chip of integrated circuits. Manufacturers will still charge 10% to maintain these new, highly reliable machines. But if they continue to offer service for your antique, it will cost a bundle. It's a lot like trying to maintain an expensive stereo tuner with vacuum tubes that you purchased in the 1950s. It's simply cheaper to buy a new tuner than to put up with the problems of maintaining the old one. Entire tuners can be purchased today for the price of one old vacuum tube—and some of those tubes just aren't available at any price.

How long will that system last? Here's how to plan. *Financially,* get the system behind you in 5 years. If the numbers don't look right on a 3–5-year basis, the numbers aren't right. *Functionally*, plan for 10 years. The programming (software) that you use should be adequate for a decade. Five to six years from now you'll be reluctant to put money into your old system, but you don't want to be pushed to buy a new one because things aren't working. If those programs (software) still grind out data in a reliable fashion, you can continue to use your old system until there is another jump in price–performance ratio. Many physicians have been forced to replace their old system because it was failing, only to see later that if they had been able to wait just 1 more year, major improvements in performance would have been possible for the same money.

In this chapter, as we examine your goals for your office computer, we will plan 10 years ahead. You'll probably trade in your equipment in 5–7 years, but we will plan so that you'll still be reasonably happy with it in 10 years.

Goals for the Decade

Ten years from now your computer should be doing a lot more than billing. In addition it should:

(1) Help to improve your day-to-day financial management
(2) Help to improve the services which you deliver to patients
    (a) by increasing your knowledge about the population which you treat (your market)
    (b) by facilitating communication with patients who need services (your marketing)
    (c) by helping you keep track of the myriads of details in your practice (your product)
    (d) by assisting with practice self-study (quality assurance)
(3) Generate revenues

These are minimums. Sit down with a pen and paper. Think of your specialty. What service will you provide? To whom? Where will they come from? Are you interested in systematic follow-up of your work? Do you wish to share data with colleagues? Write down a set of goals for the computer.

What data elements will it take to achieve each goal you have listed? The patient's age? Diagnosis? Weight? Risk factors? In general, if those data elements can be recorded on the system now, programming to support your goal will be possible in the future. In other words you can look at the computer's *future* uses by looking at *today*'s data elements.

Data Elements to Support Your Goals

Financial data elements should be sufficient not only to get your bills out and to handle third-party billing, but to improve financial decisions in your office. Most of these data elements are addressed in Chapter 3, but a few are mentioned here. It should be possible to associate each charge with a *cost center* so that the computer can be used to monitor new charges generated through the purchase of new pieces of equipment or employment of additional personnel.

(1) Family numbers should be available so that bills to all family members can be printed and mailed together. Family numbers will also let you consider entire families when analyzing bad debt experience or clinical problems.
(2) A number of data elements which do not relate directly to billing can be used to improve the services which you deliver to your patients.
   (a) You can increase your knowledge of the patients you treat by using a patient profile which includes patient address (in addition to provider's address), date of birth, sex, employer, occupation, education, genetic data, and financial class. Profiles can be made of age distributions in the practice, level of education, and occupational and genetic risk. These factors will be useful in making decisions about services which you encourage in the areas of prevention, special education, and outreach.
   (b) Date of entry into the practice, date of last visit, address, source of referral, employer, occupation, income, primary physician, and consulting physician can be used to track where patients come from in your office, why they are referred, and how long they remain with you. The data base can be analyzed to determine who leaves your practice and for what reason.
   (c) Patient address, home phone, business phone, emergency contact, and risk profile can all be used to facilitate communication with patients by printing out letters, mailing labels, or lists of phone numbers for patients who require special follow-up. Those data elements can also be utilized to contact patients who have failed to obtain services which you recommend.
(3) Problem lists, medication records, and simple outcome measures (such as weight, blood pressure, and visual acuity) can be used to facilitate practice self-analysis by selecting charts for audit, discussion, or action.

Summary

The central message of this chapter is that in selecting a computer it is important to plan for a period of 5–10 years. Before you select an office system, sit down and decide what you will want to do with the computer in 5–10 years. Once you have established goals for your system, determine what data elements will be needed to accomplish those goals.

The inventory of data elements in this chapter can be used to assess systems which are available. Several apparently simple systems will remain simple. Others make provision for collection of data elements which lay the groundwork for future analysis programming. If the data collection potential is there, then it is quite likely that the machine will serve you well.

Chapter 3

The Accounting Module

John H. Hoskins

Objectives

Describe the content of the core program in all office billing programs.

Introduction

Information processed in the practice extends beyond medical information to include nonmedical information about the patient, about the practice as a business, about the environment in which the practice functions, and about the doctor's preferred operating style.

When these information streams are formally structured and entered into a computer system, they make up what systems specialists call a management information system. The management information system for a practice may include only a very few elements, or a great many.

The subset of the management information system which includes the business plan, financial control device, scorekeeper, and planning tool for a practice is called the financial management system. It encompasses budgets (the business plan), the accounting module (the control device), a wide range of management reports (scorekeeping), and projections used in planning.

The accounting module is, therefore, that subset of the financial management system concerned with *financial control*, and its constituent programs for managing billing, receivables, and other functions are subsets of accounting.

The manager of a large clinic uses many such programs, and this system can become elaborate. In smaller practices, many parts of the system can be informal because the managing doctor is intimately familiar with the operating details.

The accounting module, however, is much the same for any practice. It requires the same sequence of steps to keep track of patient billing, no matter how many zeros are added to the totals. All that changes as a practice gets bigger is the number of people involved, the size (and cost) of the equipment, and the way the resulting information is presented.

Accounting activities, especially billing and collecting for services, use significant amounts of doctor and staff time in every practice. Computer-based systems developed to control these activities offer substantial savings in time and cost, but the benefits of systems with similar specifications vary widely depending on the quality of the system and its appropriateness for a particular practice.

The key to good program specification and selection is clarity about the objectives a machine is expected to contribute to, and about the functions it will perform.

Objectives

Computerizing practice accounting involves cost, and we expect there will be improvements which offset that cost. There are four objectives which cost-effective computer accounting should help achieve:

(1) Improving stability and continuity in the practice's daily business operations, and preventing or detecting mistakes and dishonesty
(2) Improving office life because doctors and staff spend less time doing routine accounting chores
(3) Increasing net practice income by reducing labor and outside service costs (especially accounting), by accelerating receivables and reducing lost charges, and by producing useful management reports which would be too expensive if done by hand
(4) Locking in good accounting and business practice

Functions

The simplest way to describe the functions of the accounting module is to construct it, function by function, beginning with billing and receivables control for three reasons.

(1) This function uses more staff time than any other accounting function. Its improvement is, therefore, most likely to save money and time, and improve the quality of practice work life. In particular, because a small number of information items are used repetitively, the reduction in routine data entry time is significant.

(2) Much accounting in a medical practice looks like that of any small business, but the billing process must be custom-tailored to a particular practice. Many data items are highly specific to a specialty. (There is no need, for example, for a computer in an ophthalmology practice to house detailed lists of obstetrical procedures; they would consume storage, slow processing, and never be used.) For this reason, programs for medical billing are done successfully only by doctors with computer skills and by specialized vendors whose local staff can set up a system correctly matched to practice needs. They are a demanding test of a system vendor.

(3) The billing and receivables program can produce as by-products a wide assortment of figures which are extremely useful in planning and managing a practice.

The smallest and least expensive practice computers do *only* billing and receivables. Once this function is in place, additional functions can be added up to the capacity of the machine, the availability of programs, or the desires of the managing doctor.

[Note: The volume of material needed to detail even the simplest clerical procedure always makes the procedure look forbiddingly complex. To understand this phenomenon, try writing directions for a simple *clinical* procedure (removing a suture) to be done in your absence by a bright but untrained assistant, specifying exhaustively the tools, steps, and contingencies. The billing procedure is not complex, but it has numerous sequential tasks, and we recommend a demonstration to accompany this description as the best way to understand the mechanics.]

I. Billing and Receivables

The fundamental business unit in practice is the doctor–patient encounter which generates a billable procedure. Each time this occurs (perhaps 20 times a day per doctor in a general practice; 20–40 times a month in a surgical practice), the following information must be assembled:

(1) The patient's identity and correct billing information, including insurance information if insured
(2) The doctor's identity and fee schedule
(3) The identity of all relevant third-party insurers, and claims and coverage details
(4) The diagnosis or complaint which generated a procedure, and some details about it
(5) The procedures performed, their standard charges, and the billable charges to patient or insurer
(6) The date and place of service

Reference Files

Much of this information can be entered *once* into reference files on the machine, updated thereafter only when it changes, and used repetitively without recopying, at a substantial reduction in physician and clerical time and error. Listed below are the necessary reference files, and the information items each must contain.

Patient/Guarantor Reference File

1   Patient record number
2–4   Patient name (last, first, middle initial)
5   Date of birth
6   Sex
7–10 Address and telephone number
11   Social Security number
12   Date entered in system (initial visit)
13   Relationship to guarantor (e.g., family)
14   Guarantor number (if the patient is not the guarantor, need guarantor name, address, etc., in guarantor file)
15   Primary insurance company
16   Group or ID number
17   Certificate number
18   Patient relationship to insured (if there is a second insurance company, repeat 15–17)
19+   Account controls:
      (a) bill insurance only
      (b) suspense: hold statements, insurance
      (c) in collection
      (d) no dunning
20   Billing detail, each entry containing:
      (a) date and place of service
      (b) identity of doctor
      (c) diagnosis information
      (d) procedure information
      (e) charge and billing information if nonstandard

Doctor Reference File

1   Doctor code
2   Doctor name
3–6   Address and telephone number
4   IRS number
5   Social Security number
6   Provider number
7   Medicare supplier code

8   License number
9   Specialty
10   Facility (if more than one)
11   Bill rate and factor amounts

Referring Physician Reference File

1   Name
2–5 Address and telephone number
6   Hospital or practice
7   Billing discounts/bill type

Insurance Company Reference File

1   Insurance company number
2   Insurance name
3+   Address(es) and telephone number

Diagnosis Reference File

1 Diagnosis access code (ICDA or equivalent)
2 Diagnosis description

Service Description Reference File

1   Service access code (CPT-4, RVS, or equivalent)
2   Service description
3   Category (department, specialty)
4   Standard fee
5+   Profile fees

Claim Form/Billing Form Reference Files

Instructs machine in formatting standard claim forms and standard patient billheads. One file per form.

Message Reference File

Contains numbered messages to be printed (optionally) on bills, e.g., 30–60–90 days overdue notices and the like.

Transaction Files

When an entry is made once to bill a service or enter a payment or adjustment, two things will happen:

(1) The daily transaction files for charges, payments, and adjustments are updated.

(2) The appropriate patient/guarantor file is updated.

Here are the "transaction" files used during the day for entering billing and receivables data from the terminal. Note that when the operator receives a charge ticket from the doctor after a visit, she must only enter about 40 characters of information from the form in order to have a complete, up-to-date patient record for billing at time-of-service or thereafter.

Patient Charges Journal

1 Patient number
2 Charge ticket number
3 Service date (may be automatically entered)
4 ICDA access code(s)
5 Service code(s)
6 Service fee(s) (standard or profile)
7 Doctor number
8 Place of service (office, hospital)
9 Optional text for insurance form completion

Patient Receipts Journal

1 Payment date
2 Patient number
3 Amount received
4 Form of receipt (cash, check, other)
5 Source of receipt (insurance type, etc.)
6 Charge paid (see notes)

Adjustment Journal

The Adjustment Journal collects by patient all debit/credit adjustments, insurance adjustments, and wrong-patient entries each day as an audit trail. There are four adjustments types permitted: credits, debits, refunds, and write-offs. In properly structured systems, all charge reversals are made by making a new entry exactly the same as the original entry except that the dollar figure is the negative of that in the original. A clear audit trail is thereby presented.

Reports and Information Look-Up

The information entered into the machine is used for different purposes on one of five time schedules.

1. Programs Which Operate in Real Time

Real time refers to machine use at need during each working day.

Produce charge ticket: Many commercial programs include a program which produces a charge ticket on demand. These are requested by patient number at the beginning of each working day, using the appointment calendar as a reference. They include patient identification and may include a notation of the patient's account balance and any overdue charges. They accompany the medical record to the appointment.

Enter charges in the Patient Charges Journal: The patient is asked to return the charge ticket completed by the doctor to the billing desk. The new charge information can be entered immediately, and a time-of-service bill produced if desired.

At some convenient time during the day, the operator enters transactions from hospital visits, laboratory and X-ray work, and other nonappointment procedures. There are various ways of structuring this process to avoid oversight and error. The most common, which we suggest and assume in this description, is to structure the system so that:

(1) *Every* billable transaction is entered on a charge ticket.
(2) If charge tickets are preprinted, they are numbered in sequential order, and every ticket number must be accounted for or actively voided.
(3) If charge tickets are machine-generated, the machine numbers them sequentially as requested. In this option, certain large entries such as hospital billings for the day may be entered under one number by a special subroutine.

[Note: As a contribution to reducing the paper storage problem, we note that charge tickets are *not* an integral part of the audit trail in any accounting system, manual or machine. They are simply an efficient way to communicate information from provider to bookkeeper. We suggest keeping them for 30 or 60 days after the close of the month they were generated, then discarding if no problems with the month's series are outstanding.]

Enter payments in the Patient Receipts Journal: At convenient times during the day, the operator enters all payments received, segregates checks, and batches them for daily deposit.

Enter adjustments to Patient Accounts: For billing and payment entries.

File interrogation and update: Because patient account records are always current, they can be accessed or printed out on demand to answer questions about accounts, and for other purposes. Ordinarily, any data file on the machine can be examined on demand, and the reference files updated as needed.

2. Programs Which Are Run Daily

Having collected all information to the end of the current business day, the machine is ready to assemble virtually any combination of the data fields in any of the files and display them as a report. Certain standard reports and processes are generally accepted as essential.

Daily closings: A set of programs which update the daily journal files, edit for inconsistencies and missing data and report these events, and prepare stored data for printed reports.

(1) *Charge Journal*: a complete list of all charges entered into the system aggregated by clinical service (doctor, hospital).
(2) *Receipts Journal*: which produces a daily receipts summary.
(3) *Adjustments Journal*: as above.
(4) *Bank Deposit*: a by-product of the Receipts Journal.
(5) *Missing Charge Ticket Numbers*: an important audit control. *Backup*: a set of procedures which verify that the machine has successfully closed and backed up all files used during the day. Essential to maintain the integrity of the system.

3. Cycle Billing

A major advantage of machine billing is that the billing operation can be distributed over the month. A very common schedule in small practices, for example, is to retain the monthly cycle for direct patient billing, but to do insurance billing weekly on a designated day. This greatly accelerates insurance receivables.

In the case study reported later in this section, for example, a solo practice bills about $150,000 per year in insurance billings. Weekly insurance billing reduced the average age of insurance receivables from 60 to 80 days to 15 days. There is therefore about $15,000 extra cash available to the practice, and the interest income on this cash is sufficient to pay for the maintenance of the computer.

In larger clinics, cycle billing is frequently extended to direct patient accounts, for the same reasons and with the same effect. An appropriate schedule for any practice derives from a combination of information about efficient work schedules for the staff, and the financial results of various cycles. The best pattern should be worked out with financial counsel before a machine is specified to ensure that the machine selected can perform billing on the desired cycle.

When cycle billing is used, those reports and machine operations described below as monthly which apply to receivables control will be performed more often than monthly, once for each cycle. Summary programs will be run monthly as before to provide consistent financial reporting.

4. Programs Which Run Monthly

Patient direct billing: Once a month a procedure is run which examines all patient accounts and prints a direct bill for all accounts with a nonzero balance. Referring to codes called "flags" in the patient master files, the machine automatically prints on each patient bill the appropriate message (payment instructions, "insurance not billed," account overdue notices) selected from the message master file.

Depending on the machine and printer selected, production of patient bills will proceed at a rate of 4–25 bills per minute.

Insurance forms: The insurance forms module allows for manual entry of information not normally a part of the patient's account record (e.g., disability, onset, doctor's text notes, lab, referring physician) and prepare a completed form on demand. This allows preparation of forms while the patient waits, or on a weekly cycle.

Accounts receivable summary and detail: A summary by payment source of all amounts billed in the cycle and month. A second section of this report may produce detailed billings by patient as a worksheet in processing bills to be mailed.

Revenue summary and detail: Summary by income source of all payments received during the month. Often combined with:

Credit trial balance: Verifies payments received against account balances as proof of the revenue reporting cycle.

Aged trial balance: Detail and summary by payment source of accounts with balances 30–60–90 days overdue. This report flags patient and insurance accounts which should be examined for inclusion in the collection process. Note that this is a working document, not an administrative report. The normal forms of 30–60–90-day account aging reports, which produce "percentage of collection" figures for these periods, do not produce accurate management figures. This is discussed more fully below under "Things to Watch for in Billing and Receivables."

Missing charge tickets: This report should be worked down monthly until nothing is left to establish that all charges are being promptly and properly entered into the system.

Collection letters: These are included here in their logical place. In most systems, however, the collection letter cycle is a separate program and procedure requiring manual reentry of patients and amounts selected for processing. If collection notices and letters are integrated with the production of patient direct bills, a single process may do both jobs. The best choice depends heavily on the practice's chosen approach to collection of overdue accounts.

5. Programs for Year-End Closing

The monthly summary reports, less their detail by patient and charge, are run at the end of the practice's fiscal year to produce summary figures for annual financial reports.

Administrative Reports from Billing and Receivables Data

The eight reports above (bills and insurance, receivables and payments analysis, and charge ticket control) are sufficient to control patient revenue. As a by-product of the process, other reports are optionally available which are highly desirable for management purposes—fee setting, doctor's productivity and compensation analysis, and practice analysis and development. These and their uses are described in the Administrative Module.

Things to Watch for in Billing and Receivables Programs

There are a number of ways to save money in constructing a Billing and Receivables program, and some rather fine points in accounting which make the difference between a smoothly working system and one which creates extra risk and work. Here are some major points.

(1) *Backup.* At the end of every business day there should be two copies of the complete, up-to-date contents of machine-stored data. In small machines this may require extra hardware and will certainly require extra storage media (disks, floppy disks, tapes). Older systems, designed when storage was more costly, may back up only the *changes* made in the files since the last complete backup, supposed to be done weekly or less often. The theory of this is that in the unlikely event of trouble, the up-to-date files can be reconstructed. Experience shows they are unlikely to be reconstructed right.

(2) *Account Aging.* Most systems produce aging tables constructed according to standard 30–60–90 day formulas. These produce the right information if, and only if, the billings each month are approximately constant (which they may not be, for example, during and after vacation).

(3) *Purging.* In a practice which sees many patients on a regular or hospital basis, the detailed billing information in file grows rapidly. Because vendors use the "number of patients" a machine will handle as a sales argument, and because that number is largely a function of the length of time billing detail is kept on the machine, there is a tendency to get rid of the detail and keep only the current balance as often as possible. Two principles are suggested:

(a) A system should *never* purge billing detail automatically, at the end of a month or billing cycle closing. Purging billing detail should only be done when the operator asks for it and specifies the date or specific files (e.g., before some date) to be removed.

(b) Detail should be kept long enough (a month, a year, forever) so that the bookkeeper seldom gets a call questioning a bill, only to find that the information has been purged and a search of paper records is required. These calls are time-consuming, irritate the patient and bookkeeper, and create poor patient relations. Before a system is selected, the bookkeeper should determine the required detail retention period.

(4) *Posting Receipts*. There are three levels at which systems will allow payments to be posted to patient accounts. The simplest way is simply to credit dollars. If patients both pay directly and have insurance coverage, dollar crediting is unacceptable. It makes the analysis of experience by payor impossible, e.g., how much does Blue Cross owe the practice, and what is the difference between standard fees and profile fees collected? This is essential management information.

The second level allows crediting by source of payment, which will handle most circumstances. The third level, which handles all cases, allows a hand assignment of a payment to a particular charge line in the patient account file.

(5) *Paper Wads*. Computer systems have a habit of turning out reports in enormous piles of paper, every month after a request is made once. This isn't necessary for control. Reports should not be printed more often than needed, nor should they contain any information not needed to answer the question they were prepared to answer. Many of the administrative reports will fit nicely on one or two pieces of letter-size paper. You almost never need, for example, a list of patients, although almost every system will print one.

II. Adding to the Accounting System

Three additional functions will transform the Billing/Receivables program into a rather complete cash accounting module. They are:

General Ledger/Accounts Payable
Payroll
Budget

The Accounts Payable function allows the machine to write checks and to keep track of payments as well as receipts. Because it is ordinarily installed as a part of a General Ledger System and works closely with it, the two are discussed together. Payroll systems keep track of payments, deductions, and withholding amounts due; completes various quarterly and annual payroll tax reports; and writes payroll checks. Payroll can be installed either independently or interconnected to the General Ledger. The latter is more effective, but more complex. We consider payroll separately.

Limits to Accounting Systems

Our discussion of extensions beyond Billing and Receivables is not detailed. The reason is that these options produce escalating requirements for expertise within the practice, and how much is done depends on internal staff capability. It is useful to divide accounting systems into three groups according to the difficulty of supervising them:

(1) Billing/Receivables. Any practice can manage these; they are actually easier to control than a pegboard system because of internal checks and automatic totalling.

(2) *Cash* accounting systems which include Payroll and General Ledger/ Accounts Payable. These require about the same skill level as the same items done by hand internally. In many small practices, however, most tax returns, financial statements, and the like are done by the outside accounting firm. The skill level required to perform these internally may be higher than present staff affords. The computer does not add the complexity; the additional duties add a knowledge requirement. Formal bookkeeping training will be required. In larger practices, the skill level required is that of full-charge bookkeeper.

(3) We have not considered extensions of the accounting system into accrual books, keeping depreciation schedules, and other elements of a complete corporate set of books. There are no medical office systems which reach this level of complexity on a "canned system" basis because there is always a considerable amount of special design involved. The in-house participation of a manager with accounting training will be required to develop or modify such a system to meet the requirements of an individual practice.

In order to extend the accounting system to include General Ledger/ Accounts Payable, we will need the following reference files:

Chart of Accounts

1 Numbers for various income and expense types
2 Description of types

Vendor File (Optional)

1   Vendor number
2   Vendor name
3–6 Vendor address and telephone number

Budget

A budget is compiled if the system supports budget-to-actual comparison including Payroll.

Employee Reference File

1   Employee name
2   Address

3   Social Security number
4   Date of birth and hire date
5   Data about rate and frequency of pay
6   Number of exemptions
7   Data about deductions for benefits and other items (other than income tax and FICA, which are calculated)
8+ Other information required to make the payroll system come out right. This can be quite complex.

Tax Tables

For federal, state, and local taxes

We will need transaction files.

Employee Transactions

Carrying hours worked and other information entered by pay period

Invoice Register

1 Date
2 Vendor
3 Description of payment
4 Amount
5 General Ledger expense type (if not automatic)
6 Check number (assigned serially by the system)

We should be able to produce at appropriate intervals at least the following reports:

Case disbursements and check registers

Check reconciliation listings

General ledger income and expense statements

Balance sheets

Payroll and vendor checks

Employee quarterly and yearly payroll registers

Employee deduction registers

Tax returns and federal deposit registers

Things to Watch for

Chart of Accounts: A file listing all of the income and expense, asset, and liability accounts for the practice, and their descriptions. There is a de facto standard for accounting systems for medical practices, published by the Center for Research in Ambulatory Health Care Administration (CRAHCA) as *Practical Financial Management for Medical Groups* (Schafer et al., CRAHCA, Medical Group Management Association, 4101 E. Louisiana Avenue, Denver, CO 80222; $50) At least 10 national vendors of office computer systems have advised the MGMA that their systems will accommodate charts of account consistent with this standard.

The standard covers structural matters in accounting: charts of account, budgeting, cost accounting, and financial reporting. Its use is valuable because (1) all the necessary categories of income and expense are represented and (2) they are defined in a standard, well-defined terminology which makes assignment of items easy, which facilitates comparison with other practices. We recommend that in-house systems which extend to General Ledger conform to this standard.

III. Quality of Practice Life: A Case Study

We have suggested that in addition to possible financial benefits, the installation of a Billing and Receivables system tends to improve the quality of life in the office. To explain this assertion, here is the report of one solo practitioner.

The practice is in internal medicine, specializing in nephrology. A typical week has about 120 office visits, heavily weighted toward exams, and 40 hospital visits. Much EKG and laboratory work is done in the office. Total staff is two full-time and one part-time assistants, all of whom do all the jobs: laboratory, bloods, cardiology, reception, and billing. About 350–400 statements are mailed each month in the billing cycle, and about 200 insurance forms. The purpose of the computer was primarily to reduce the agony of billing, not to reduce cost.

Before: Every Sunday, the doctor dragged home the week's insurance forms and spent 3–4 hours completing them. Some weeks he couldn't face it. Saturday before billing day each month, after rounds, he spent 4 hours going through the box of pegboard cards, adjusting bills, pasting on three flavors of past-due notices, and assigning cards to three stacks; "bill," "don't bill," and "send to collection agency."

On billing day, an assistant spent 2 hours at the copy machine running billls. There were then two stacks; cards and bills. Two more hours were required to refile cards in alphabetical order. Bills were checked against the ledger and the ledger totalled. Sometimes two cards stuck together and one was overlooked. The doctor was asked to help find out why the pegboard didn't balance. Fold, stuff, and mail bills.

After: Run monthly billing program. Bills are itemized, overudue notices automatically printed. Process takes 2–3 hours, during which the assistant can be doing something else. Doctor makes one pass through printed bills, separating into three stacks; "send," "don't send" (Blue Cross contracts will pay, courtesy, adjustments), and "collection." Time: 45 minutes.

Insurance forms are largely completed by machine, and go out weekly on Thursdays, a maximum of 6 days after service. Average age of insurance receivables dropped from 60–80 days to 15 days. The practice bills about $150,000 per year to insurance; there is therefore about $15,000 extra cash available. In 1980 it was earning 14% or $2100 per year, which pays for the machine service contract.

The practice originally used self-mailer bills, which cost 30 cents each, but found them a pain. They switched to ordinary flat statements. Using these, the total cost of billing supplies is about $200 a year, saving the expensive costs of pegboard forms, copier liquid, and toner.

Assistants like the system; the quality of life is better. It is easy to check receipts; the machine makes a daily deposit slip which must agree with cash or something's wrong. The doctor likes it; many useful management reports are available on demand, such as procedures for period, amount charged, and average collected. Costs are known, so the reports are a big help in keeping the fee schedule within the reasonable profitability range. Saturday and Sunday afternoons are free to spend with the doctor's family.

Cost: The machine cost $21,000, software $6000, initial supplies $500, installation zero. The doctor put $7000 down, and took a 36-month reducing loan at 1% above prime. Payments are $579/month principal, plus declining interest. The practice took the regular investment tax credit and $470/month depreciation.

The machine will probably last 10 years; it will be obsolete, but the job won't change, so it will still perform. Recoveries against cost were minimal; a possible 1% improvement in receivables, but these were already high. The doctor's overall assessment of the change is an important improvement in the general character and conduct of the practice for the cost of any moderately expensive piece of practice equipment.

Training: Everyone went to an evening course for raw beginners at the local technical college, and learned a little vocabulary, which cured most of the assistants' apprehensions. It was important to have picked a course NOT taught by someone interested in explaining how complicated it all is; it isn't. The doctor then went on vacation while assistants had 4 days of hands-on vendor training. All new staff have picked it up hand-to-hand.

Startup: The machine and pegboard ran in duplicate for the first 6 functioning weeks (they opted to bring patients on the machine as they had new visits—there are other ways). For 2 weeks they used the pegboard to bill, with the machine in parallel. For 4 weeks they used the machine with the pegboard in parallel. Then they used the pegboard only for patients with no new entries. At the end of the fourth month, there were only four bills left in the old system, all long-overdue collection items, so the pegboard was closed.

IV. Possible Figures for Accounting Module Section

1. B/R: Patient/Guarantor Reference File Contents
2. B/R: Doctor Reference File Contents
3. B/R: Referring Physician Reference File Contents
4. B/R: Insurance Company Reference File Contents
5. B/R: Diagnosis Reference File Contents

6. B/R: Service Description Reference File Contents
7. B/R: Claim Form/Billing Form Reference File Contents
8. B/R: Message Reference File Contents
9. B/R: Patient Charges Journal—Sample Lines
10. B/R: Patient Receipts Journal—Sample Lines

11. B/R: Adjustment Journal—Sample Lines
12. B/R: Sample Header for Machine-Generated Charge Ticket
13. B/R: Sample Machine-Generated Bank Deposit
14. B/R: Sample Lines—Missing Charge Ticket Report
15. B/R: Sample Page: Accounts Receivable Summary and Detail

16. B/R: Sample Page: Aged Trial Balance
17. B/R: Sample Page: Revenue Summary and Detail
18. B/R: Sample Page: Credit Trial Balance
19. B/R: Sample Page: Payment Pattern Table for Receivables

20. Budget: Sample 1-Page Practice Budget
21. Budget: Sample Page: Budget to Actual
22. GL: Sample Page: General Ledger Chart of Accounts
23. GL: Sample Page: Invoice Register
24. GL: Sample Page: Summary Income and Expense (Cash)

Part II

Introduction to Administrative Systems

Chapter 4

The Administration Module

John H. Hoskins

Objectives

Present important administrative uses for information originally entered for billing purposes.

Computers in Administration

Where the Administration Module Fits

There are resources besides dollars which are used in every practice. The use of these resources must be planned, scheduled, administered, and reviewed. They are time, facilities, equipment, and supplies.

Strategic Planning

Decisions arise irregularly in every practice about the overall stock and disposition of resources. For example: whether to add staff or additional physicians; whether to purchase specific equipment; whether to change office hours, open a satellite clinic, or make an arrangement with a hospital.

The character of these decisions is that they are commonly expensive and create long-term obligations. If the decision-maker has access to enough data about the practice as it presently *is*, discussion takes on an air of reason which improves the possibility of making the right decisions, and decreases the likelihood that expensive decisions will be made in thalmic response to short-term crises. Computer support of administrative systems—scheduling, inventory, correspondence, and reports—should therefore be evaluated both in terms of its contribution to daily activities, and in terms of its contribution to overall practice planning.

Patient Demographics: Administrative Use of Accounting Data

The first step taken by consultants in practice development is to convince the managing doctor that intuitively knowing who the patients are (which everyone does) is not the same thing as knowing analytically. Analytical information is essential to thoughtful practice management. From the information entered into a billing system we can obtain (by machine; these counts are too expensive to generate by hand) the following information.

Total Active Patients

Once a year, a program should run which gives the total count of patient files which have charge entries dated within the past 12 and 24 months. The purpose of obtaining this figure is to compare it to the same figure 1, 2, and 3 years ago, as a measure of practice growth.

Total New Patients

Once a year, a program should run which produces the count of patients whose date of first visit is within the past 12 months. The purpose of obtaining this figure is to compare it to the same figure 1, 2, and 3 years ago, as a measure of practice health, and to compare it by year with the inactive patient count described next.

Total Inactive Patients

Once a year a program should run which counts and *lists* all patients who have not been seen in the past 36 months. The purposes are (1) to compare this figure with the new patient count as a measure of practice health and (2) optionally, to provide a list of patients whose account files (if paid current) may be purged from the machine. Purged accounts will be set up anew if the patient returns later.

Patient Count by Zip Code

An annual run of patients sorted by zip code of home address, with the results (number and percent) plotted on a zip code zone map available from the post office is an accurate and inexpensive way to define the service area of the practice.

Patient Distribution by Demographic Characteristics

Tables of patient age, sex, occupation, and other demographics, cross-referenced by diagnosis, give one view of the nature of the practice and secular changes in it. A common pattern for older doctors, for example, is a decline in the number of new patients, an increase in the average age of patients, and a secular shift in diagnoses toward those characteristic of older people.

Patient Count by Primary Insurer

Practices which draw heavily for their patients on employees of large local employers will want to keep one eye on these employers. Possible catastrophes include plant closings, layoffs, or major change in insurance coverage (such as a new Health Maintenance Organization in the area) which might attract employees. The first step is to ask, "who pays the bill?"

A parallel measure is *Dollar Volume by Source of Payment*, including Medicare/Medicaid, state, and other insurers.

Procedure Frequency and Dollar Volume by Procedure

This program describes the nature of the practice, and may suggest greater or lesser efforts to attract certain kinds of business.

Source and Volume of Referrals

By referring doctor, (1) number and kind of procedure per year, and (2) dollar volume. The purpose is to determine who is supporting the practice and compare this with 1, 2, and 3 years ago. Are referrals still coming? Is the practice heavily dependent on one or a few referring doctors?

Production by Doctor

In various forms this information may detail the procedures, dollar volume billed, net dollar volume collected, and discounts from standard fees for each doctor in a practice. These figures, especially when compared to previous time periods, have multiple uses. They measure individual and relative workload, and relative dollar productivity (which may impact on profit distribution plans).

A small number of these studies—typically those related to production—may be wanted on a scheduled basis. Most well-constructed accounting modules offer a selection of "canned" reports which can be produced either automatically at each billing cycle or on demand. Less common, but invaluable when needed, is the ability to cross-tabulate virtually *any* set of variables contained in the patient and accounting files.

It is difficult to draw the line between clinical and administrative uses of this kind. For example, a program which records diagnoses, but does not record current medications, may still be useful if a medication used in the practice is recalled. The doctor is likely to know with certainty the diagnoses for which he might prescribe the drug; a printout of patients with these diagnoses and a call to the pharmacy can short-circuit a week-long task of scanning charts.

Patient Scheduling, Follow-Up, and Recall

Programs for patient scheduling and recall are a problem of sizable proportions. The reason is that, although they provide useful time management tools—hospital lists, morning appointment lists, lists of patient charts to be pulled and ready, lists of patients scheduled for recall—a fully developed scheduling program must operate in "real time," which in this case means they must be responsive to changes in the daily schedule because of emergencies, running late, and rescheduled appointments.

The program needed to response to such changes is complex and requires both the time and attention of the appointments secretary to keep it up to the minute. It may require more time than it is worth to keep it up to the minute during the day.

[There are exceptions. In Radiology practices, where the requirement for efficiency is to schedule patients *and* staff *and* multiple facilities concurrently, the scheduling secretary is an important part of the management team, and a machine schedule can make a real contribution. Some specialized and very elaborate programs for this purpose exist.]

On the other hand, the use of a machine program for patient follow-up and recall may contribute to both good medicine (not dropping the ball on patients who medically *should* be seen on a schedule) and financial growth (by reliably bringing patients back on a schedule). Such programs are based on a machine-stored perpetual calendar and allow entry at time of visit for follow-up calls, cards, doctor actions required, and the like.

Many practices which use machine scheduling have found the following compromise practical:

(1) During the day, all new appointments and cancellations for following days (at least a year ahead) are entered, as are follow-ups and recalls according to predesigned schedules for various classes of patients. Changes in today's schedule are not entered.
(2) Each morning, printed lists are distributed to the doctors, the appointments secretary, and others who may need them. These include appointments for each doctor, hospital lists, and lists of special actions such as recall notices to be mailed, phone follow-ups, and the like. These lists are updated during the day *by hand* as circumstances change.

The result is a scheduling operation which begins each day in good order, but does not impose the requirement of printing new lists and distributing them several times a day. Because the daily changes usually have minimum impact on forward scheduling, review of the daily schedules over a period of time gives good information about time utilization of doctors and staff.

Inventory

If the practice stocks many items in quantity, good inventory programs are available for some (not all) office systems. The purposes of installing such a program are (1) to prevent runouts and provide some control against pilferage and (2) to reduce the amount of money tied up in inventory to the minimum consistent with (1). Money on the shelves in supplies earns nothing; it could be earning 10–15% in a money market fund.

Word Processing

Word processing on the office computer is obtained by adding a software package which includes an "Editor." There is some confusion about the cost-effectiveness of doing this, the result in part of equipment vendors having taken the position that it will help everybody, which is not the case.

The question has been studied exhaustively by industry and government agencies, with results which are based on the underlying fact that a skilled typist does dictation, transcription, and copying from longhand of *original material* faster and more accurately on a typewriter than on any computer terminal. The computer comes into its own when there is a considerable body of material which is typed repetitively with no changes or only moderate amounts of change.

(1) If the practice correspondence and other typing is primarily original material, the most cost-effective device is an electric typewriter.
(2) If a typist or transcriptionist is employed half-time (or one-half a full-time equivalent) in the practice, it is cost-effective to spend an additional $1500 to have a typewriter with a "self-correcting" feature.
(3) If 8 hours a week or more are spent reproducing by typewriter standard paragraphs of text (such as the "boiler plate" of referral letters) and the amount of such material is small enough to fit in memory, it may be cost-effective to spend a further $1500–$2500 to add memory to the typewriter.
(4) If, and only if, there is substantial copying or minor editing of "canned" material, the favored choice is word processing on the computer. Such material is likely to be found in:
Welcome-to-practice letters
Referral letters

Preoperative and postoperative reports
Signature-on-file letters
Insurance company letters
Referral report letters
Name and address labels for records, laboratory specimens and reports, and the like.

A survey of uses should be taken before committing to word processing. Moreover, it should be absolutely determined that the word processing program will not be appreciably slowed down by other programs running on the machine. This can easily happen in "multitasking" computers, where more than one task at a time may run under a computer-monitored schedule. The word processor, which has a person watching it, is the one that will suffer by interruption for another task.

Part III

Introduction to Health Care Delivery Systems

The clinical applications of computers to Health Care Delivery aids the physician in the patient care management of his or her practice. Chapters 5 through 13 address nine basic areas where the computer can assist in improving patient care.

The physician will not utilize all of these areas today or even tomorrow. However, if physicians are aware of their existence, future planning can be accomplished within the initial office study and systems analysis.

Chapter **5**

How Computers Can Help in Patient Care and Practice (Health Care Delivery)

Byron B. Oberst

Objectives

Give a general overview of how computers can be applied to health care delivery in an office setting.

Delivery of Health Care

The potential of the computer to aid health care delivery in private practice is just beginning to develop. Hospital information systems are basically administrative-accounting systems with secondary applications to health care delivery. If computers are to be clinically useful, it is important that physicians become involved in the adaptation of the computer to private practice and patient care. If physicians do not define and spell out their patient care needs, computer software specialists will unwittingly develop programs which are far from satisfactory and not very practical.

There are at least nine areas of patient health care delivery that lend themselves to computer applications. Physicians need to become aware of these areas and define each of them so that software specialists know what is desired and needed.

The component parts of a health care record system include the following:

1. Initial data base
2. Active working record for daily use including:
   a. Accounting
   b. Administrative data
   c. Health care delivery
      (1) Professional continuing education
      (2) History gathering and data input
      (3) Medical history synopsis with updates
      (4) Surveillance and patient recall.
      (5) Quality control and personal care audits
      (6) Summary health record for patient use
      (7) Statistical data collection
      (8) Outcome of treatment
      (9) Other

Physicians need to review their own office practices and begin to evolve a reasonable structure and consistent method in patient care. The heart of this structure is a dynamic, active, uniform office record system with a practical system of coding. Though physicians would like to have more descriptive and useful coding systems for diagnoses and procedures than the International Classification of Disease System and the Current Procedure Terminology (CPT), these two systems are all that are practically available and already are widely used in hospital and third party payee systems. In pragmatic terms, physicians must devise ways of using these current particular tools. Systems such as the Systematized Nomenclature of Medicine (SNOMED) are too cumbersome for office use where the bottom line is so important.

Given this practice structure which makes recall and surveillance of all types possible, the opportunities for computer application are unlimited. The nine areas of application will be briefly discussed in this chapter.

1. Time planning is a vital concern for most physicians. Time commitments involve continuing medical education, practice management, hospital service, medical society and community committee work, personal family requirements, state relicensure, and medical specialty society recertification. These and similar time needs which require documentation are obvious choices for computerization. By accruing and analyzing a year's time utilization in these areas, a physician can plan and allocate his or her time resources in a more efficient manner. Some of these time demands are the following:

   (1) Educational seminars
   (2) Authoring of articles and publications
   (3) Lectures
   (4) Postgraduate courses
   (5) Conventions
   (6) Independent study and self-instruction

(7) Reading of monographs or books
(8) Audiovisual study
(9) Videotapes
(10) Professional meetings
(11) Professional committees
(12) Practice requirements
(13) Community organizations
(14) Office management
(15) Preparation for seminars and meetings
(16) Medical teaching
(17) Patient care review
(18) Self-assessment and self-evaluation
(19) Cardiopulmonary resuscitation certification

2. A physician would find very useful a brief summary of a patient's medical record which contains the following items:

(1) Anthropometric measurements of height, weight, blood pressure, pulse, temperature, respiration rate, head and chest circumference (especially in infants and young children), and skin-fold thickness
(2) Demographic data
(3) Medication list
(4) In-office laboratory data
(5) Out-of-office laboratory data and consultation reports
(6) Listing of various visit types and procedures performed
(7) Problem list and the status of each problem
  a. Active problems
  b. Inactive problems
  c. Potential problems
(8) Immunizations
(9) Brief salient family history data
(10) Brief salient patient history data
(11) Adverse reactions to medications and other substances
(12) Patient and family functional capacity and status

With the above data, the physician has an enormous capacity to define and intervene in the management of patient health problems. With facile update, change, and correction, the Medical Synopsis becomes a working, dynamic document rather than an archive repository.

The Medical Synopsis can become the patient's own personalized index of medical needs including the following areas:

(1) Indicates need of specific disease flow sheets.
(2) Indentifies patient recall needs
(3) Records types of care rendered
(4) Records surveillance needs

(5) Assists in collection and retrieval of personalized data and statistics
(6) Facilitates ease of transmittal of essential health data to other uses
(7) Facilitates audit of quality of care and examination of treatment costs

3. There are a number of methods available to obtain medical history data including programmed history forms, use of cathode-ray tube (CRT) terminals, and processing of data by optical scanning equipment. These techniques can summarize large amounts of information in a usable manner with the aid of the computer.

4. Surveillance of defined patient care needs is an area where the computer can be appropriately and efficiently applied. Monitoring of immunizations, supervision of needed return visits for chronic conditions (e.g., diabetes, cardiac problems, convulsive disorders, and learning disabilities), monitoring certain medications, recall of patients who miss follow-up visits, and determination of patient treatment compliance are all areas where computers are helpful. Using this type of computer surveillance can help to avoid potential medicolegal liability.

5. There are many areas of patient care in medical practice which lend themselves to quality control. The computer can monitor completeness of treatment, abnormal laboratory results, conformity to predetermined algorithms for standard of care, functional outcomes, and document improved follow-up of patient care, determine how well a particular medication is working in a certain disease state, indicate areas of possible need for physician self-education, and relate specific medical literature references to unusual or exotic disease states.

6. The computer can offer major assistance to the physician in his or her own continuing medical education program. He can utilize it to define treatments used and analyze outcomes. A physician can readily determine what types of patients are seen and what age groups are involved. For example, physicians provide many different types of care such as:

(1) Emergency care
(2) Hospital care
(3) Acute episodic care
(4) Consultative care
(5) Convalescent care
(6) Prescheduled care
    (a) Health supervision
    (b) Chronic care
    (c) After care

By determining his or her own activity mix, disease entities, and various types of care rendered, a physician can readily determine areas in which to seek appropriate medical education and technical training. Chapters 6 and 12 address these issues in more detail.

7. Patient, parent, and family education is discussed in Chapter 11. Medicine of tomorrow will be highly involved in all types and methods of

patient education. Computers have an important potential role in this area.

8. There is a wealth of health information obtained by census surveys which can be compared to individual office practices by relating census track information to zip code areas. The physician can use these methods to determine which areas of his or her practice are adequate and can identify geographic indicators suggesting intervention such as pockets of inadequately immunized patients.

9. There will be many other computer applications to health care delivery developing in the near future such as transmittal of data within regional health care areas, networks, and other larger systems.

With interested and knowledgeable physicians teaming with innovative software specialists, the horizons for health care applications of computers are almost limitless.

Chapter 6

How Physician Professional Education and Development Can Be Enhanced by Computers

Byron B. Oberst

Objectives

Present techniques to log data and utilize office computer systems to enhance the physician's professional environment.

Introduction

Good patient care and treatment programs are dependent upon the physician keeping abreast of many small packages of changing knowledge, changing technology, and changing demands on his or her time.

The computer can be a useful tool to aid in many applications in this area.

(1) It can be used to accumulate and document hours of categories I through V of formal continuing education for the AMA Recognition Award, state relicensure, and specialty society requirements for recertification or maintenance of specialty society membership.

(2) It can be used to document practice management time so that the time used by this management activity can be allocated to the cost of each activity of the practice. Time dedicated to management can also be documented for use in practice income distribution.

(3) The computer can be used to document professional committee work in hospitals, and time dedicated to medical staff work, and local, state and specialty medical organizations. These activities are very time-consuming in every practice. Efforts in each area should be documented to help look at practice priorities and family priority planning. It is useful to determine if the time spent in these activities enhances your practice. Do they obtain new patient referrals? Do the experiences broaden one's own medical vision and vistas? Do they provide more knowledge? These are questions that should be examined as the busy physician tries to prioritize demands on his or her time and energy.

(4) The computer can log the various types of educational materials utilized by the physician and document the time devoted to each educational method (see Table 6-1):

(a) The use of journals and which specific journal
(b) The use of audio tapes services and which ones
(c) The attendance at seminars, round tables, and workshops
(d) The preparation time for presentation at meetings, teaching ward rounds, journal clubs and similar activities
(e) Other

(5) An analysis of patient records in a horizontal manner over time develops a total picture of the longitudinal aspects of a disease state or treatment program.

(6) An analysis of patient records in a vertical manner regarding medical specifics and the outcome of a particular disease state or treatment program can be accomplished.

Table 6-1   Summary of The Professional Activities of B. B. Oberst., M.D.; Clarkson Hospital Data—1980

Type	Category	Credits
CME Accredited Sponsorship	I	116
CME nonaccredited Sponsorship	II	16
Medical Teaching	III	34.5
AMA Recognition Award, Papers, Publications, Books, Exhibits	IV	0
Nonsupervised Individual CME	V	165
Other Meritorious Experiences	VI	35
Nonaccredited Activities Office Management		301.2
Total Hours		667.70

(7) An analysis of accumulated statistics can be performed regarding various disease entities by specific subsets of patients or various age groupings. These analyses may indicate areas which the physician may wish to emphasize in his program of continuing education.

(8) The computer may be used to compare a single patient, a group of patients, or a treatment program in a physician's own practice with a prepared algorithm. The algorithm may be of the physician's own planning or from an outside source. It may contain expected treatment methods, treatment outcome, expected patient compliance, cost analyses, or personnel activities within the physician's office setting. For example, in my office, at periodic intervals, the nurses' aides conduct a manual audit on various age brackets of Health Supervision Visits (Annual Health Reviews) to determine if all designated data is being collected as programmed. This type of audit uses management by exception to quickly identify any major deficiencies. This audit enhances awareness of my responsibilities to my patients for quality care, my personnel's regard for my proper supervision of their activities, and my pride and awareness of a good job being done for my patients. This technique also identifies errors which need to be corrected. A spin-off benefit is that I seek other areas of application within the office setting, read material on patient care, or attend discussions on practice management with a closer focus on my own deficiencies.

The following list of office personnel activities can be delegated by the physician but he or she remains responsible for the proper execution of these duties.

(1) Nurse–patient history interviews
(2) Nurse–patient educational activities
(3) Speech screening
(4) Hearing screening
(5) Vision screening
(6) Nurses' patient observation notes
(7) Specialized educational activities
(8) Audiometer testing
(9) Growth charts
(10) Developmental testing
(11) Reading testing
(12) Spelling testing
(13) Word recognition testing
(14) Patient body dominance
(15) Developmental drawings of geometric figures; man, woman, and house
(16) Specialized adolescent measurements such as biocrominal and biocrestal diameters

Chapter 7

The Medical Record Summary, Contents, and Utilization

Gretchen Murphy

Objectives

Identify the role of a medical synopsis for office practice.

Describe a computerized medical synopsis for office practice.

Introduction

The primary goal of the health care system is to provide health care services in an accesible, cost-effective manner. Today's private practitioner is expected to provide more complex services at the primary care level. Physicians maintain and extend diagnostic and referral services. Billing, scheduling, patient records, insurance reporting, health screening, and marketing are practice needs that must be met. Computers offer considerable assistance in all of these areas.

There is an interest in computerizing information in the patient record to develop a medical summary capable of supporting ongoing clinical needs of patients. Table 7-1 shows that 43.5% of the data handling in physicians' offices is directed to problem definition and updating, medical summaries, follow-ups, and surveillance.[1] A well-defined, computerized medical synopsis can provide a major contribution to physicians in patient information handling.

Table 7-1 Percentage Allocations Between Medical and Administrative Data

Medical Recording—Total		67.5%
File index retrieval	.5	
Physicians' commentary such as definition and update of problems, medical summaries, follow-ups, or surveillance	43.5	
Ancillaries (vital signs and measurements, histories)	5.0	
Physicians' orders—prescriptions, reports to GPs, referrals	8.0	
Laboratory reports—preparation and/or interpretation	4.0	
X-ray report processing	2.5	
Health risk appraisal processing	4.0	
Patient Scheduling—Total		9.0
Appointment processing	5.0	
Registration (demographic and billing data)	4.0	
Clinical and Patient Education-Oriented Applications—Total		8.0
Administration Applications—Total		15.5
Bill processing	5.0	
Accounting	4.0	
Payroll	1.5	
Personnel	1.5	
General reporting and statistics	3.5	
		100.0%

Source: Jan F. Brandejs and Graham C. Pace. *Physician's Primer on Computers*, Lexington Books, D.C. Heath and Company, Lexington, Ma., 1980. Reprinted with permission.

Computers in Ambulatory Care

The role of computers in ambulatory care has been well recognized. An extensive study of the role and use of the Automated Ambulatory Medical Record Systems (AAMRS) in 175 specific clinical sites was performed in 1975 and updated in 1981. A major premise of these systems was the capturing of patient data at the primary care level. The following design characteristics[2] were included in these systems.

Medical Services

(1) Patient profiles—a concise summary of a patient's medical status
(2) Patient surveillance reports—information used in preventive care and management of chronic disease
(3) Time-oriented flow charts or other standard formats for data presentation showing a sequence of medical data
(4) Computer-generated encounter forms
(5) Progress notes in text form

(6) Medical histories
(7) Database searches—data retrieval to serve information needs of training, research, and medical audit

Administrative Services

(1) Accounts receivable and billing
(2) Third-party claims: eligibility determination and claims preparation
(3) Reports for management and supporting agencies
(4) Input to other accounting systems

Other Services Provided by the Systems

(1) Appointment scheduling
(2) Registration
(3) Medical record accession for hospital-based clinics
(4) Medical education
(5) Data for research

Patient Data Entered in the Record

Patient information was captured to provide part or all of the patient record to the physician via computer screen displays and/or hardcopy printouts on a day-to-day basis.

The amount of medical data entered varied from practically none to the complete medical record. Data entered consisted of both free text and code. In general, coded input was used for physical examination results and lab results. Text was used for chief complaint, problem lists, medications, and notes.[2]

This medical data often comprised a medical synopsis for the providers. The data was entered from an encounter form that was filled out by the physician. The length of the encounter form varied from a simple one-page document to complex multiple-page documents with computer-generated forms in use at some sites. A few systems accepted direct input through a CRT terminal by physicians. The most significant contribution of these systems was the direct incorporation of the patient medical information as a primary goal in the initial development.

Current Status

In 1983, several of the earlier promising systems became operational and underwent transfer from a prototype demonstration in the research setting to use in the commercial market. The leader in commercial availability of an in-depth medical record system is COSTAR. Other systems have continued to be, or are in the process of becoming, commercially available. These include AUTOMED, TMR, and RMIS. Each needs to be evaluated to see how it meets the needs of an individual practice.[2]

Today, additional computer-supported, patient information systems are needed to facilitate primary care delivery. For many physicians, the more highly developed systems are not practical at this point. The need for an effective financial application is their first consideration. Improved patient information handling is next, and a computerized medical synopsis is important as well.

This chapter explores the role and use of a medical synopsis or the medical data portion of such systems and identifies a method to investigate current alternatives. It is possible for practitioners to begin with simple applications and develop more complex models over time. However, this requires some investigation and evaluation. The following questions identify and illustrate some key factors to be considered.

(1) What is a medical synopsis? What are the objectives of a computerized medical synopsis?
(2) How can the primary care physician go about determining the contents of an appropriate computerized medical synopsis for use in his or her practice? How can existing models offer ideas and suggestions? Is there a recommended standard data set for the synopsis?
(3) How is patient information captured for input and processing?
(4) What kind of outputs should be designed?
(5) How can the medical synopsis serve as an information base for other, nonclinical uses?
(6) Are there current resources that can assist the physician in this area?

What Is a Medical Synopsis?

A medical synopsis is a summary form of patient information. It can range from a simple abstract that contains brief demographic data and a diagnosis to a complete summary of patient visits including treatments and results. A computerized medical synopsis can provide a minimum medical database for primary care patients in summary form. It should contain sufficient information so that the need for the entire paper record is reduced or eliminated. The objectives of a medical synopsis are:

(1) To provide a summarized form of the patient's record to facilitate physician use of the patient information
(2) To document previous and ongoing visits to track problems identified and treated
(3) To provide a succinct information summary to expedite evaluation during clinic visits and for use in referrals

(4) To provide a medication list to monitor individual drug profiles
(5) To guarantee a summarized and easily maintained medical problem list in the record of each patient
(6) To provide information for disease and procedure coding for billing purposes
(7) To provide analytical research data for the physician

How Can the Primary Care Physician Decide on the Contents of a Medical Synopsis Appropriate for the Office Practice?

An effective medical synopsis can be developed or acquired by assessing and evaluating patient data needs for the practice. Such information needs can (generally) be determined in the following way.

First, physicians can examine some general patient encounter forms which satisfy visit data collection needs and accommodate billing. Tables 7-2 and 7-3 are illustrations of physician office or other ambulatory care encounter forms now in use.[3,4]

Note that the following minimum data appears:

(1) Name   (5) Responsible party or insurance carrier
(2) Address, zip   (6) Current phone
(3) Identifier   (7) Reason for visit/complaint
(4) Date of birth

This is basic patient information that is also provided in the commercial billing systems that are generally available. However, the encounter form samples also provide data for a synopsis or medical summary which could be appropriate for general medical record-keeping. Sample forms like these can be reviewed to ascertain readability and information flow. Practice staff can offer valuable assistance at this point as they must extract patient information for a variety of uses. They also should have an understanding of the information flow within the office.

Supplemental encounter forms could be used to provide additional data collection for special needs and/or referrals. For example, a page 2 form can be used to expand the information on the patient where additional treatments are prescribed. There are many existing encounter forms which address the immediate reason for the visit. Many commercial computer service companies now include a list of problems or diagnoses in their programs. Others have introduced a medication list to the package they market. Physician and medical computer user groups such as the MUMPS Users Group publish information and serve as resources for novices.

Second, once an encounter form has been identified and assessed for the basic information it contains, a method of linking one visit to another should

Table 7-2 Emergency Department Encounter Form

```
09/09/81              THE JOHNS HOPKINS HOSPITAL CORE RECORD (EMERGENCY DEPARTMENT)      Z0001
--------------------------------------------------------------------------------
ADDRESS:  ████████              TIME IN: 01:00 PM   HISTORY NUMBER:  ██████
CITY: BALTIMORE                                     NAME:  ██████████
STATE: MD      ZIP: 21218    PHONE: ████             DATE: 09/08/81
--------------------------------------------------------------------------------
  ONSET OF ILLNESS-ACCIDENT        : COMPLAINT
      DATE    : TIME
      /   /   :        AM  PM     :
  RACE : SEX : DATE OF BIRTH : MARITAL :   RELIGION    :   NEXT OF KIN-NAME    :   RELATION
   B  :  F  :    12/27/27    :            :             :                      :
      FATHER'S NAME          : MOTHER'S MAIDEN NAME  :   NEXT OF KIN ADDRESS
                             :                       :
  CLINIC NO.    :   TRANSMITTAL NO.  : FINANCIAL CODE : HOW ARRIVED  -WALK  -BUS  : ACCOMPANIED BY
                :                    :      SRC       :   -CAR  -TAXI  -AMB#      :
                :                    :                :   -POL#       -HEL#       :
                        PERSON RESPONSIBLE FOR BILL / EMPLOYER
  NAME / EMPLOYER
--------------------------------------------------------------------------------
                              BLUE CROSS PATIENTS
      MEMBERSHIP NUMBER      :  POLICYHOLDER NAME        RELATION
                            :  OUT STATE ADDRESS        PLAN #
--------------------------------------------------------------------------------
      MEDICARE NUMBER        :        MEDICAID NUMBER             DATES
                            :         ██████       02/76 TO 12/81
================================================================================
: DOCTOR'S INSTRUCTIONS           : NURSE'S INSTRUCTIONS                         :
...
: DIAGNOSIS:                      :
: PHYSICIAN'S SIGNATURE:          : NURSE'S SIGNATURE:                           :
--------------------------------------------------------------------------------
I ACKNOWLEDGE THAT THE ABOVE INSTRUCTIONS HAVE BEEN EXPLAINED TO ME TO MY SATISFACTION.
    PATIENT'S SIGNATURE:
    IF THERE ARE ANY QUESTIONS REGARDING YOUR TREATMENT, OUR STAFF WILL ASSIST YOU - CALL 955-2280.
    BRING THIS SLIP WITH YOU IF YOU RETURN TO THE EMERGENCY ROOM.
    ALWAYS BRING YOUR HOSPITAL PLATE WITH YOU WHEN COMING TO THE HOSPITAL.
*** DISPOSITION ***
( ) PRN                              ( ) ADMISSION
( ) ER/PCC RETURN VISIT              ( ) JHH CLINIC VISIT
( ) LAB & XRAY TEST                  ( ) OTHER
( ) COMMUNICABLE DISEASE REPORTED    ( ) NOT TREATED
              *** ORIGINAL MUST BE FILED WITH MEDICAL RECORDS ***
```

Source: E. McColligan, B. Blum, and C. Brunn, An Automated Core Medical Record System for Ambulatory Care, Figure 2, *Computers in Ambulatory Medicine*, Proceedings of the Joint Conference of the Society for Computer Medicine and the Society for Advanced Medical Systems, 1981, p. 76. Reprinted with permission.

be required so that summary data can be produced. While such summaries are established features of the more highly developed systems, more limited applications may only store the current visit or the current and previous visit. Unless this need is specifically addressed, it may not be provided. Using a common patient number would be one way of providing this link. To illustrate, the patient number could be used in addressing segments of computer stored encounters on floppy disks. The linkage component should provide continuity to chronic problems and treatment modalities.

Third, the encounter form should be analyzed to see if it includes an appropriate demographic and clinical profile of the patient. One model for this is shown in Table 7-4. It was developed by the Society for Computer Medical Standards Committee. It includes uniform data set for ambulatory records, which is an important element in securing comparative information from individual and group practice activities.[4,5]

Fourth, the encounter form should be checked to see that provision is made for IDC-9 and CPT coding in the synopsis. While these codes are used primarily for billing purposes at the present time, they have the capability to provide analytical and research data for the physician. Retrieval by diagnosis/problem would be valuable both to the physician and to the epidemiological surveyor.

Fifth, information transfer of the synopsis for referral and for patient use should be considered. To illustrate, if a medical synopsis was stored on word processing equipment, it could be played out for referrals for transfer to another physician. It would also be helpful in information transfer to hospitals, nursing homes, and home health care programs. In the 1980s, we can expect to see an increase in the use of prehospitalization work-ups performed in physician offices, recorded via CRT, and transmitted to hospitals through teleprocessing.

Sixth, other physicians and office practice staff who have had experiences with computerized systems should be polled. The low-cost microcomputer market offers several alternatives. Using existing software such as Visifile and database management packages, experience can be gained in creating and working with medical synopsis files.

How Is the Patient Information Captured for Input and Processing?

Components of the synopsis are captured via the encounter form. The synopsis is then computer-generated with the preprinted information identifying the patient and provider and the medical data that has been collected on the encounter form. Patient records are usually processed before the day of the visit with the synopsis report placed in the chart or replacing the chart. During the patient visit, an encounter form is completed and any changes are

Table 7-3   Office Encounter Form and Attending Physician's Statement

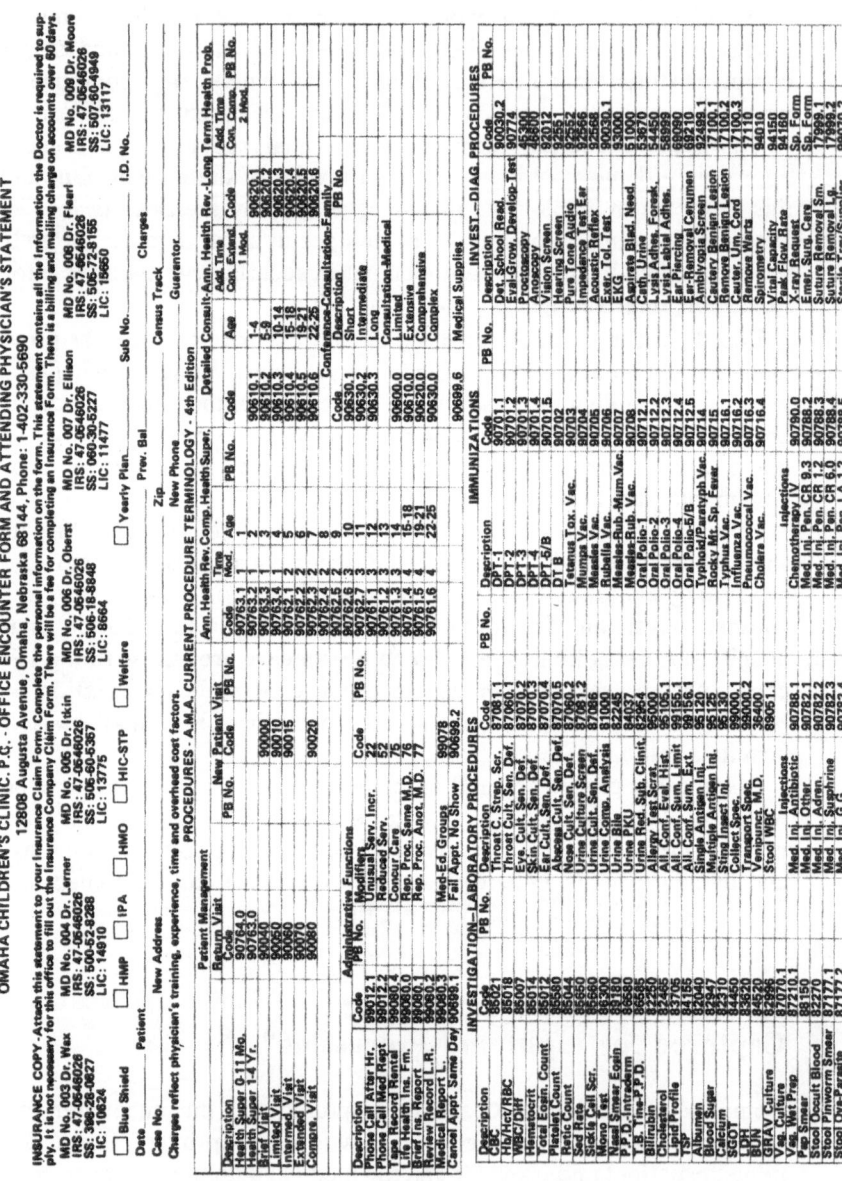

Reason For Visit ___ HSV No ___ ☐ Laceration 879.7 Location ___ ☐ Health Super V20.2 ☐ Med Check Date of Birth ___ ☐ Sick Check Male ☐ Female ☐ ☐ Consultation Education Level ___ ☐ Referral MD No.

ICD.9 CM DIAGNOSTIC CODES – CPHA 1978

Description	Code	PB No.	Description	Code	PB No.	Description	Code	PB No.
Gastroenteritis	008.0		Serous Otitis	381.4		Acne	706.1	
Diarrhea	009.3		Dys. Eust. Tube + Vent. Tubes	381.81		Seborrhea	706.2	
Chicken Pox	052.9		Sensorineural Deaf	389.18		Urticaria – Hives	708.9	
Herpes Simplex	054.9		Rheumatic Fever	390.0		Collagen Disease NEC	710.9	
Rubella	056.9		Hypertension	401.9		Synovitis	727.0	
Inf. Mono	075.0		Cardiomyopathies	425.4		Osteochondritis	732.7	
Viral Warts	078.1		Dis. Heart Rhythm	427.9		Arrest Bone Develop	733.91	
Viral Infection NOS	079.9		Sinusitis	461.9		Tibial Torsion	736.89	
Tinea Unspec.	110.9		Acute Pharyngitis	462.0		Scoliosis	737.30	
Thrush	112.0		Acute Tonsillitis	463.0		Con. Hear Dis. NEC	746.9	
Pinworms	127.4		Acute Tracheitis	464.1		High Arch Palate	750.26	
Helminthiasis NOS	128.9		Croup	464.4		Undescended Testicle	752.5	
Pediculosis	132.9		Acute URI	465.9		Metatarsus Adductus	754.53	
Scabies	133.0		Bronchiolitis	466.1		Congenital Anomaly Unspec.	759.9	
Hemangioma	228.00		Hypertrophy – T & A	474.10		Sm. for Gestational Age	764.2	
Simple Goiter	240.0		Hypertrophy – Adenoids	474.12		Post Term – Post Mature	766.2	
Hypothyroidism NOS	244.9		Pneumonia	486.0		AB History Distress Nb.	768.9	
Diabetes	250.01		Influenza	487.1		Hemolytic Dis. Nb.	773.2	
Hypoglycemia	251.2		Bronchitis	490.0		Neonatal Jaundice	774.6	
Delayed Puberty	259.0		Asthma	493.9		Feed Problems Nb.	779.3	
Precocious Puberty	259.1		Teething Syndrome	520.7		Other Spec. Nb. Condition	779.8	
Lactose Intolerance	271.3		Dental Caries	521.0		Syncope	780.2	
Unspec. Dis. Lipid Metab.	272.9		Impacted Teeth	524.3		Febrile Convulsions	780.3	
Rickets	276.2		Stomatitis	528.0		Sleep Disturbance	780.5	
Obesity	278.0		Inguinal Hernia	540.9		Fever Und. Origin	780.6	
Anemia – Iron Defic.	280.9		Umbilical Hernia	550.90		Malaise – Fatigue	780.7	
Anemia – Unspec.	285.9		Constipation	564.0		Anorexia	783.0	
Coagulation Defects	286.9		Irritable Colon	564.1		Failure to Thrive	783.4	
Allergic Purpura	287.0		Hepatitis	573.3		Constitutional Smallness	V72.8	
Thrombocytopenia Unspec.	287.5		Hematuria	599.7		Headache	784.0	
Enuresis	307.6		Hydrocele	603.9		Speech Disturbance	784.5	
Encopresis	307.7		Vaginitis – Vulvitis	616.10		Funct. Heart Murmur	785.2	
Adjustment Reaction	309.9		Dysmenorrhea / Pain Menses	625.3		Enlarged Lymph Nodes	785.6	
ADD-MBD-LD	314.0		Irregular Menses Cycle	626.4		Abdominal Pain	789.0	
Slow Learner	319.0		Cellulitis	682.9		Frax. Unspec. Bone	829.0	
Meningitis	322.9		Acute Lymphadenitis	683.0		Concussion	850.9	
Cerebral Palsy	343.9		Impetigo	684.0		Superficial Injury	918.9	
Convulsive Dis. – Epilepsy	345.1		Pyoderma	686.1		Adverse Reaction – Medicine	995.2	
Migraine	346.9		Granuloma Umbilicus	686.1		Health Supervision	V20.2	
Encephalopathy Unspec.	348.3		Urinary Tract Infection	599.0		Single Newborn	V30.0	
Myopia	367.1		Atopic Dermatitis – Eczema	691.8		Premature Infant	765.1	
Amblyopia	368.00		Dermatitis	692.9		Twin Birth	V31.0	
Conjunctivitis Unspec.	372.30		Allergy Food	693.1		Fam. Hx. Cancer	V17.9	
Obstruction Tear Duct	375.55					Fam. Hx. Card – Vas. Dis.	V17.4	
Strabismus	378.9					Fam. Hx. Allergic Dis.	V19.6	
Otitis Externa	380.10					One Parent Family	V61.9	
Impacted Cerumen	380.4							

Patient Measurements: HT, WT, BP, HR, RR, Temp, Head, Chest, Skinfold, Other

98 Child Past History
99 Family History

Pt. Health Code Status

No.	Description
1	Episodic
2	Long Term
3	Chronic
4	Improving
5	Inactive
6	In Evaluation
7	Potential
14	
15	Adverse Effect
16	Case Closed
17	Surg. Repair
18	Resolved
19	Case Reopened
20	
21	Acute

No.	
1	Spec. Care Prog.
2	Lowfat Diet
3	Bland Diet
4	Reduce Diet
5	Ulcer Diet
6	Special Diet
7	Lact. Free Diet
8	Diabetic Diet
9	High Prot. Diet
10	Elim. Diet
11	Rotation Diet
12	Gluten F. Diet
13	General Diet
14	Environ Cont.
15	All. Hygiene
16	Sinusitis Rou.
17	Phys. Therapy
18	Psychotherapy
19	Group Therapy
	Phys. Appl.

Functional Assessment Code

1	2	3	4	5	6	7	8
Medication	PB No.	Date		Name-Brand or Generic Title			

Change PB No. ___ to PB No. ___
Change PB No. ___ to PB No. ___

MD Signature ___

Guided Growth ☐ Printed Mater'l ☐ Aud-Vis Aids ☐ Lectures ☐ Other ☐ Return Visit ☐ Return Phone ☐ Return LMD No. ☐

Patient - Parent - Education   Disposition

Literature - References

PB No. ___ Date ___   Batch ___

Table 7-4 Summarized Health Profile

Demographic/administrative	Medical
Identification	DX/Problem list (for each)
Unique identification #	Name (codable)
Name (surname, first, middle)	Modifier (noncoded)
Former names (maiden, etc.)	Estimated date onset
Birth date	Date first dx'ed
Birthplace (city, county, state,	Cause or etiology (primary,
nation, etc.)	secondary to, etc.)
Race	Parameters (defined) to be
Sex	monitored and frequency
Social Security #	Most current date of verification,
Twin marker	modification, etc.
Communication	Current status (active, inactive,
Mailing address (residence)	preventative, etc.)
Demographic address (county	Current therapy
code, census tract, etc.)	Source of information
Telephone (business, home)	Parameter results
Emergency notification	Prior date of verification,
(relationship, address)	modification, etc.
Next of kin (relationship, address)	Status
Social	Therapy
Marital status	Source
Religion	Parameter results
Education	Critical med. information (allergies,
Occupation	sensitivities)
Previous occupations	Confidential med. information
Number of people in household	Present health status
Head of household (name,	(Physical, emotional)
relationship to patient)	Scheduling
Family linkage designation	Date last update
Financial	Pending laboratory studies
Employer	Scheduled dx procedures
Payment type	Scheduled therapies
Payment source and address	Scheduled reassessments
Financial history as per third	
party requirements (including	
financial ID#)	
Information about primary	
responsible provider	
Name	
Unique identification #	
Address, phone	
Type (M.D., D.O., dentist, nurse,	
SSW, etc.)	
Speciality	
Date initialized	
Date of last update	

recorded. For the initial clinic visit, the encounter form can be used to create the first synopsis. Subsequent visits are usually recorded on the synopsis printout itself or on a new encounter form.

Data entry can be performed by the office staff using the information recorded during the visit. Alternative forms require one to check a box, make a number, mark a space, or write or dictate a note. Each visit date may be generated by a computer program for all synopses printed for the day in question, or it may be entered directly on the CRT by the clerk at the time each day's visit encounters are entered.

Only key data necessary for the patient synopsis is entered and stored on the computer. All other data may be filed in the paper record. The problems and medication lists should be created and updated by the provider at the time of each visit.

Some computer systems utilize database software technology so the record can be consolidated and each visit entered into a central database under the name and record number. In group practices, confidentiality can be protected with the use of user codes and passwords to access particular data.

What Kind of Outputs Should Be Used?

The synopsis or summary is the most important product in the practice. What information will be needed from one visit to another? What information needs to be summarized? Tables 7-5 and 7-6 illustrate two examples of a medical synopsis, also called a patient profile or summary.[6,7]

It is frequently useful for physicians to start with the outputs so that they fully consider the visit information needs. Consideration of alternative encounter forms must be keyed to a clear understanding of the required clinical data.

How Can the Medical Synopsis Serve as a Base for Other Information Needs?

A medical synopsis can provide a base for other information needs. If properly designed, the information capability derived from the data elements included could be a resource for the following:

(1) Practice profiles by patient age, demographics, insurance, and referrals. Private carriers and Medicare play a significant role in patient activity. These can be traced periodically to assist the physician in determining existing practice characteristics and plan for changes.
(2) Problem and procedure profiles can be constructed from indexing the disease and procedure codes used for billing. This is currently done by commercial computing services for nursing homes. It is a low-cost batch process.

Table 7-5   Medical Record Copy for Permanent Chart Entry*

```
*********
*PROBLEX* Patient ID No. 100033  Physician No. 00099     Home ████████
*INQUIRY* Name ████████             ████████   Sex M    Work 000-000-0000
* CHART * Addr ████████████████                          Phar ████████████
*  COPY * Last Visit 02/22/81  Freq 02  Last RX 02/22/81  Birth ████████
*********

PROBLEM LIST  (X = INACTIVE)    C A   THERAPY          DOSAGE    FREQ    PROB

01 ASHD                         3 COUMADIN           5 MG      AS DIR      7
02 ANGINA PECTORIS              3 LANOXIN            0.25 MG   1 QD        5
03 DIABETES MELLITUS            2 ISORDIL            2.5 MG    1SL AC&HS   2
04 GRAVES' DISEASE            R 1 INDERAL            40 MG     2 BID       2
05 ATRIAL FIBRILLATION          2 NITROGLYCERIN      GR 1/150  1 SL PRN    2
06 THYROTOXIC EXOPHTHALMOS       2 SINEQUAN           50 MG     1 HS       11
07 PULMONARY EMBOLISM           1
08 NEUTROPENIA                  2
09 COLON CARCINOMA            S X
11 DEPRESSION                   2


DRUG ALLERGY-INTOLER (I)    DRUG ALLERGY-INTOLER (I)    DRUG ALLERGY-INTOLER (I)

PENICILLIN

RX REFILL




Page 1                         Medical Report                        7/31/81
                    Harwood Medical Associates,S.C.
                         7400 Harwood Avenue
                         Wauwatosa, WI 53213
                           414-771-5300

    Patient:  ████████   ████████                        Sex M
                                                   Birth date ████████
              ████████        ████████            Last visit 02/22/81
              Home Phone ████████████             Weight 164
              Record # 100033                     BP 140/070


    Problems                      ICD-9-CM  Assessment

    ASHD                          414.0     PROBLEM CONTROL
    ANGINA PECTORIS               413.9     PROBLEM CONTROL
    DIABETES MELLITUS             250.0     BORDERLINE CONTROL
    GRAVES' DISEASE               242.0     IDEAL CONTROL
    ATRIAL FIBRILLATION           427.3     BORDERLINE CONTROL
    THYROTOXIC EXOPHTHALMOS        376.21    BORDERLINE CONTROL
    PULMONARY EMBOLISM            415.1     IDEAL CONTROL
    NEUTROPENIA                   288.0     BORDERLINE CONTROL
    COLON CARCINOMA               153.9     INACTIVE PROBLEM
    DEPRESSION                    311       BORDERLINE CONTROL

    Therapy          Dosage     Frequency   Drug Allergy or Intolerance

    COUMADIN         5 MG       AS DIR      PENICILLIN
    LANOXIN          0.25 MG    1 QD
    ISORDIL          2.5 MG     1SL AC&HS
    INDERAL          40 MG      2 BID
    NITROGLYCERIN    GR 1/150   1 SL PRN
    SINEQUAN         50 MG      1 HS
```

*This medical record copy is produced for permanent chart entry after each patient visit. Space is allowed for written notes. A second copy is edited by the physician and used by medical transcription for updating the computer record.
Source: Daniel J. Forward, *Development of a Computer-Assisted Problem-Oriented Medical Record System for Office Use,* Computer Applications in Medical Care, Proceedings of the Fifth Annual Symposium, Copyright © IEEE Computer Society, 1981, pp. 106, 109, 110, Appendix. Reprinted with permission.

Table 7-5 (*continued*)

```
Page 2          ████████    ████████             Record # 100033

Chronology

1906  BIRTH
1912  MEASLES
1913  MUMPS
1929  T&A
1956  COLON CARCINOMA              Surgically Treated
1967  RECTAL POLYPS                Surgically Treated
1972  CARDIAC CATHETERIZATION
1973  ACUTE MI                     Hospitalized
1977  GRAVES' DISEASE              Radiation RX
1978  PNEUMONIA                    Hospitalized


Intercurrent illnesses / Followup Parameters
Date         Contact   Problem              Plan / Parameter

11/22/79     OC        VIRAL URI            SYMPTOMATIC
1/12/80      PH        PROSTATITIS          BACTRIM-DS
2/20/80      PH        PROSTATITIS          BACTRIM-DS
5/30/80      PH        STREP THROAT         ERYTHROMYCIN
3/20/80      OC        OTITIS MEDIA         ERYTHROMYCIN
11/10/80     PH        SINUSITIS            ERYTHROMYCIN


Comments




This information is confidential and may not be released to other
parties without written patient authorization.

                        --------------------------------M.D.
                        ████████████████████            M.D.
```

(3) Cooperative practice evaluation and assessment could be performed through screening programs applied to the medical synopsis.

(4) Practice profiles on drug therapies related to chronic problems could assist in Item 3.

Are These Current References Available to Assist the Physician in Decision-Making?

Current references of operational computerized medical record systems appropriate or adaptable for ambulatory care at the physician practice level are available in the literature. Materials are available which describe existing ambulatory care medical record systems. One such summary can be found in *Computers for the Physician's Office*.[8] The authors describe several ambulatory care models. With the exception of the COSTAR program, most are designed for an individualized setting. As previously stated, the COSTAR

Table 7-6   Confidential Patient Health Summary

```
HEALTH SUMMARY
  (07/09/81)                BETHEL
                            ***CONFIDENTIAL PATIENT DATA***

NAME: ▓▓▓▓       �▓▓      ▓           MEASUREMENTS  (MAX 5 OR 2 YEARS)
▓▓▓▓▓▓                                 DATE     WT  PCT  HT  PCT  BLD PRS
SEX: ▓▓▓                             11/24/80  120-08  29  63    8  110/076
BENEFICIARY: ▓▓▓▓                    08/20/80  111-12  17  62    8  110/064
BIRTH: ▓▓▓▓                          12/22/80  104-04  17  59   3-  102/064
PCIS ID NO:
SOCIAL SECURITY NUMBER: ▓▓▓▓         TONOMETRY READING
REGISTER NOS: AR-SU-FY-REG NO
  HNDCP. CH.   30-31-HC-000123       VISION
  MT ED H S    30-36-SA                                      UNCORR    CORR
                                     08/18/80   MT ED HC   L 20/030
H.R. NOS:     AR-SU-FY-REC NO                              R 20/030
  ANMC        30-31-01-000656
  YUKON KUSK  30-33-01-001234        IMMUNIZATIONS        ı
  MT. EDGE    30-36-01-054321        SMALL POX       12/02/70  BET S PHN #1
  BETH PHN #1 39-06-58-654321        DPT           B 08/07/74  BET S PHN #1
                                     DPT           3 07/12/71  BET S PHN #1
ACTIVE PROBLEMS                      DPT           2 04/28/70  BET S PHN #1
   YR   DATE                         DPT           1 02/13/70  BET S PHN #1
NO. ENT MOD    FACILITY   DIS PROBLEM OPV          3 07/12/71  BET S PHN #1
001 79 01/23/79 MT.ED HC  MD  COM WITH PERFORATION R OPV      2 04/28/70  BET S PHN #1
       09/18/80 N001      MD  SCHEDULE T-PLASTY    OPV        1 02/12/71  BET S PHN #1
*002 80 10/14/80 YUKON KUSK MD P/O TYPMANOPLASTY 10-10-80 TYPHOID  08/14/75  MT ED HC
003 80 01/25/80 BET PHN #1 PHN AD PERF CHRONIC OM MEASLES      09/17/71  MT ED HC
                                     RUBELLA       09/17/71  MT ED HC
INACTIVE PROBLEMS                    MUMPS         09/17/71  MT ED HC
   YR   DATE
NO. ENT MOD    FACILITY   DIS PROBLEM SKIN TESTS  (MAX 3 FOR EACH TYPE)
                                     TINE      10/20/77  BET S PHN #1  N 00
                                     TINE      11/99/74  BET S PHN #1  N 00
INPATIENT ENCOUNTERS (MAX 9 OR 2 YEARS) PPD     12/10/80  MT ED HC   UNKNOWN
10/09  10/12/80 YUKON KUSK  OTITIS MEDIA PPD     08/20/80  MT ED HC      N 00
                            RIGHT DRUMHEAD PERFORATION MONOVAC 06/01/81  BET PHN #1  N 00
                            R TYPE I TYMPANOPLASTY
                                     LAB-X/RAY RESULTS  (MAX 7 OR 2 YEARS)
OUTPATIENT AND FIELD ACTIVITIES (MAX 18 OR 2 YRS)          12/20/79
05/02/81 MT ED H C  OTO    COM AD W TM PERFORATION HEMOGLOBN   14
04/13/81 MT ED H C  OTO    COM AD
02/05/81 KWETHLUK   PHN    COM AD              REGULAR SURVEILLANCE
                           IMPETIGO                          LAST         NEXT
12/22/81 KWETHLUK   PHN    PPD GIVEN          TB TEST     06/01/81     06/01/82
                                              BP          12/22/80     12/22/81
```

Source: C. Constance Relien, Tracking and Training Tribulations of a Data System in Alaska, *Computers in Ambulatory Medicine*, Proceedings of the Joint Conference of the Society for Computer Medicine and the Society for Advanced Medical Systems, 1981. Reprinted with permission.

system is currently being marketed as a transportable package, and there are some recent adaptations now available on microcomputers. In their discussions, the authors identify some major questions that should be considered when "contemplating automated records systems." These questions also apply to the use of a medical synopsis. For many physicians, a medical synopsis may be a computer-provided summary which is part of an ongoing paper record. In other cases, it may become part of the paper records after serving billing needs. The questions raised by Zimmerman and Rector are:

(1) What are the missions and objectives of the physician's office served by the records system?
(2) For what medical and other purposes are the data used?

(3) What data are stored? In addition to patient identification and demographic data, what data are stored to describe the patient's problems or diagnoses, allergies, etc.?
(4) To what extent are the stored data to be encoded as opposed to free text?
(5) Who collects the data and who enters it into the computer?
(6) By what physical means are the data collected and entered (form, equipment, etc.)?
(7) Are the data retrieved from the computer by the physician himself, or by others? What data is used most frequently during patient visits?
(8) What are the technical facts, such as type of computer to be used and the computer language in which the application is written?
(9) What is the cost to develop or acquire the system, and what is the cost to maintain it? How are these costs divided between hardware and software? How much does the system save?
(10) Is the system potentially transferable elsewhere? If so, what are the costs and constraints?

These questions focus on the need to clearly identify "how to" methods for the interested reader.

References

1. Brandejs, J; Pace, G: *Physician's Primer on Computer—Private Practice.* Lexington Books, D.C. Heath & Company, Lexington, MA, 1979, p. 67.
2. Kuhn, IM; Wiederhold, G: The evolution of ambulatory medical record systems in the U.S. *Proceedings of the Fifth Annual Symposium on Computer Applications in Medical Care*, IEEE Computer Society, Washington, D.C., 1981.
3. Encounter Form (Children's Clinic of Omaha).
4. McColligan, E: An automated care medical record system for ambulatory care. *Computers in Ambulatory Medicine Proceedings, Joint Conference of the Society for Computer Medicine and the Society for Advanced Medical Systems*, Washington, D.C., 1981.
5. Jelovsek, FR, *et al.*: Guidelines for user access to computerized medical records. *Journal of Medical Systems*, Vol. 2, No. 3, 1978.
6. Forward, DJ: Development of a computer assisted problem oriented medical record system for office use." *Proceedings of the Fifth Annual Symposium on Computer Applications in Medical Care*, IEEE Computer Society, Washington, D.C., 1981.
7. Relien, C: Teaching and training tribulations of a data system in Alaska. *Computers in Ambulatory Medicine Proceedings, Joint Conference of the Society for Computer Medicine and the Society for Advanced Medical Systems*, Washington, D.C., 1981.
8. Zimmerman, J; Rector, A: *Computers for the Physician's Office.* Research Studies Press, Forest Grove, OR, 1978.

Chapter **8**

History-Gathering Techniques via the Computer

John M. Long

Objectives

Discuss one very specific and often beneficial aspect of an automated medical record system in a private practice office, namely, the direct entry of patient history data into a computer via a terminal.

Compare this approach with the more conventional method of history-taking, suggest when it is practical, and provide guidelines for those who wish to proceed.

Introduction

Background

It is possible, even relatively simple in today's technology, to have your patient directly enter into the computer demographic and medical history data via a computer terminal. This is done using a keyboard and screen or other direct entry devices such as a graphics tablet. Such applications are commercially available either in a separate (the so-called "stand-alone") computer application or as one component of a modular system. Since direct entry devices are used in many contexts today, most physicians have some knowledge of them. Nonetheless, it would be helpful to review some of the relevant options available for entry of data into the computer.

(1) The most common data entry method today is via a keyboard. Often the keyboard is a standard typewriter or a calculator keyboard or both. Usually a screen is attached so that the user can see what is being input.

(2) Sometimes the keyboard has a special design and/or is programmable. For example, many fast food chains with a limited number of items for sale use a keyboard with a key for each item on the menu. These devices are programmed to print the name and price of the item and to calculate and print the tax and total cost. Programmable keyboards have a wide variety of potential applications in medicine, but have received very limited use at this time.

(3) Touch sensitive screens are available whereby the user simply enters data by touching the item or option listed on the screen which is appropriate to the specific situation. For example, in history-taking, a question is flashed on the screen with various options for answers. The patient would simply touch the proper response. Some touch screens are sensitive to the touch of a special light pen.

(4) A far less common data entry device is the graphics tablet. Whatever is written on the tablet is transferred to the screen.

History-Taking Methods Using a Computer

Forms

By far the most common method today for entering a patient's history into a computer is to collect the data on a form and to later enter the data into a computer. This method is also almost universally used to enter patient data collected by the physician who uses one or more standard encounter forms as discussed and illustrated in Chapter 7. Attempts to develop systems whereby the physician directly enters data into the computer have been unsuccessful. Even the paperless doctor's office described in Chapter 14 uses a doctor encounter form which is virtually the only manually completed form in the office.

Patient-Operated Terminal

Systems which allow the patient to enter his or her history data directly into a computer have been successful but are not widely used. There are commercially available systems and in certain situations they are especially useful. For example, mass screening systems use this approach. Some physicians successfully use them in their office. Dr. Sheldon Cohen reports such a use in his private practice.[1] An assistant takes about 3 minutes to teach the patient how to use the computer terminal. Each history takes about a half hour. It is used two to four times per day and costs, with maintenance, $327 per month or about $5.00 per history. Either the touch sensitive screen or graphic tablet could be used for this approach but have not been so far as we know.

Assistant- or Nurse-Operated Terminal

A third method involves one keyboard and two CRT viewing screens with a nurse or office assistant operating the keyboard. One screen is for the patient and the other for the nurse or office assistant. The questions are asked and responses are entered by the assistant. The patient can review the data on his or her screen in order to verify its accuracy. This approach is especially valuable for collecting demographic data.

Why the Limited Use Today?

There are a number of reasons why automated history-taking is not widely used today.

(1) Limited use. In the typical private practice office, patient histories are probably not that frequent. The primary use would be for new patients along with periodic updates for others. This may be a minor activity in many offices.
(2) Limited value. Unless the automated histories are a module in a more comprehensive automated medical history system, its use is somewhat limited and, chances are, the patient can fill out a paper form with almost the same benefits.
(3) Further development is needed. Much progress has been made to improve and humanize the man–machine interface. Nonetheless, much more development work is needed before there is wide acceptance of the concept of directly entering data into a computer, especially by the physician. The trend seems almost inevitable, but is not now upon us!

Why Would One Want to Consider
History-Gathering via the Computer?

There are reasons for using automated history-gathering systems today.

(1) It has been shown that they can be made to work. Patients will use the terminal and can be quickly taught how to use them.
(2) If the history-gathering unit is a module of a more comprehensive system, it can be most valuable. This is especially true if the system has the ability to organize the data and highlight the key information for the physician and/or if the data can be fed into other modules related to billing, practice management, and patient management including patient education.

What Data Should Be Collected?

The standard questions found on paper forms are more or less directly transferred to the computerized version. Compare the information collected on the computerized version with that now collected on the existing manual system. Unless the computer version is at least as complete and, in addition, has the ability to integrate, organize, and highlight critical data it is doubtful that it will be worth much.

Some Guidelines

It would be a very unusual case where automated history-taking would be a logical first step in the automation of your office practice.

If you are automating any aspects of your office, say the financial aspects, you should also consider using an automated history-taking module if available.

The module should be integrated into the other modules of your system.

Reference

1. Cohen, S: "Experience with a computerized medical history system in private practice," pp. 121–123, *Proceedings of the Fifth Annual Symposium on Computer Applications in Medical Care*, IEEE Computer Society, Washington, D.C., 1981.

Chapter **9**

Supervising and Keeping Track of Patient Care

Gretchen Murphy

Objectives

Provide an explanation of patient surveillance and tracking in office practice.

Identify potential content of a patient surveillance system.

Examine key issues involved in determining the appropriateness of a surveillance system.

Introduction

Patient surveillance and follow-up systems can help physicians facilitate preventive health care and manage chronic health problems in their office practice. The ability of the computer to maintain updated lists which are continually modified is an ideal asset in this area and with interest in patient life-style as a primary factor in health care management, surveillance and follow-up have become increasingly important. In their discussions of the current status of computer applications in ambulatory care, Kuhn and Weiderhold included a surveillance factor in the expected benefits of computerized ambulatory care systems. Patient surveillance reports, identified as information used in preventive care and management of chronic disease, were listed as major design components in the emerging AAMRS settings.[1]

This chapter describes patient tracking as a practice management tool. Six issues are reviewed to assist physicians through an effective consideration of this tool.

(1) What is a surveillance system and how does it function?
(2) What are the benefits in acquiring such a system for office practice?
(3) What information would be tracked in surveillance and follow-up?
(4) Is it realistic to consider microcomputers to operate it?
(5) How can individual data elements for a proposed surveillance program be identified?
(6) Are there existing resources now available?

What Is It and How Does It Function?

Surveillance and patient tracking can be defined as a system established to identify and monitor specific problems, diagnoses, treatment parameters, or results as evidence in objective data review, and medication or treatment response. Surveillance can be established for individual patient care management in which ongoing visits are projected and related patient data is recorded and tracked. It can also be used to record collective results of treatment of like conditions for the purpose of increasing the physician's knowledge about his or her own practice.

The most common form of collecting data for surveillance purposes is the encounter form. For physicians who are not yet ready to use billing and encounter forms for this purpose, a brief abstract form could be adapted to meet the needs. A simple surveillance process could work in the following way.

A basic registration form completed in the physician's office can be used to register a patient into the practice. Patients can be registered either on a terminal or through a data form that is later batch processed by a computer service bureau. Basic data elements needed are:

1. Patient name
2. Address
3. Zip
4. Identifier (this may be a number that links records)
5. Date of birth
6. Responsible party or insurance carrier/number for billing
7. Current phone number
8. Social Security number
9. Reason for visit
10–13. Current medications
14–16. Other selected data elements

As a practical matter, these items should be available to the practice staff. Given these data elements, a file can be created and stored on a microcomputer using a commercial software package such as Visifile, the Data Reporter, or File Manager. Master name indices can be printed out by number and/or alphabetical lists. Such lists would provide a ready reference for practice staff. By extending the data elements to add a surveillance code, several useful things can be accomplished. For example, each patient could be coded for surveillance as follows.

Code	*Dictionary*
01	30-day follow-up (medication, blood pressure)
02	Calculation for immunization schedule and projected return visits
03	Follow-up referral to specified date (longer time periods, i.e., blood level checks)

Lists of patient names can be provided by code number for notification. Notification can be performed as an ongoing manual task by the practice staff or, in more sophisticated programs, through reminder letters that are generated as part of the system. Word processing equipment would provide considerable assistance in storing standard letters and reminder notices. Some physicians may want to use a computer service to batch process the registration and surveillance information and provide name lists and surveillance lists by code number. The practice staff could then work directly from the lists and prepare letters and notices using standard formats stored on word processing equipment. This would require less initial outlay and still yield an opportunity for working with the computer options. Then, when more automation may be needed, this experience would facilitate decision-making at the next level because basic data use will have been validated. In addition, new uses of the data may have been identified.

What Are the Benefits of Surveillance Programs to the Physicians?

Benefits of practice surveillance systems are many. Not only can the individual physician maintain cumulative data on specific patient problems, but groups of physicians can pool data for clinical investigation and analytical use.

Surveillance systems provide a method for monitoring those patients who are participating in treatment programs.

As has already been described, surveillance programs promote a systematic practice management that can result in less stress and better physician/patient support.

Cumulative use of surveillance data on chronic conditions can be valuable in assessing medication use in long-term patient management.

Cooperation with external data collection and surveillance programs will yield both clinical and community analysis that can provide significant assistance in future planning.

One example of a comprehensive practice surveillance system which records results of therapy for individual and collective use is the Cardiology system at Duke University which maintains current results of therapy for hypertension patients. These results subsequently are available to participating physicians for reference in prescribing their individual cases.

What Information Can Be Tracked in Surveillance Programs?

Patient surveillance systems can track generically for general follow-up or specifically for individual patient monitoring.

In generic tracking, all patients in the practice are tracked for routine physical examination, or, in the case of pediatrics, for immunizations.

In specific surveillance, patients are tracked according to problem or diagnosis, or medication and treatment. The following list identifies typical information monitored in these systems.

(1) Cancer therapy
(2) Diabetes treatment and response
(3) Hypertension management
(4) Blood level checks for drug therapy in epilepsy
(5) Follow-up programs for heart attack patients
(6) Prosthetic device management
(7) Pacemaker management

Other conditions could include alcoholism, blood dyscrasias, arthritis, rheumatoid conditions, postoperative conditions, and other problems identified by the individual practitioner.

Is It Realistic to Use Microcomputers for Surveillance Programs?

Microcomputers are used today as components of special purpose applications in medicine and health information; as components of office automation; and as personal tools which can be used for home and business applications. Medical research, medical data collection, storage, retrieval and manipulation, and medical decision-making are examples in which microcomputers are employed by health care professionals.

Table 9-1   Data Elements for Tracking Epilepsy Patients

Patient name
Address, zip, phone
Patient number
Diagnosis
Drug identified for monitoring—title and dosage
Dosage date of last blood level test
Results of previous blood level test
Monitoring cycle (60 day intervals)

Today microcomputers are used in physicians' offices to perform billing and accounting functions, clinical record-keeping, word processing, and inventory control.[2]

Simple surveillance applications can be developed with existing commercial software packages that are marketed for use on the personal-sized computers such as the TRS-80, APPLE, and the IBM-PC. Generally speaking, the state-of-the-art is not yet developed so that novice users can immediately use these machines to develop a particular application without a significant investment in time. This is needed to become familiar with the machine. However, if a physician is interested in experiencing the microcomputer, the outlay of approximately $3000 may well justify some time investment in the system.

Given a software database management package, students spent between 25 and 30 hours developing simple surveillance applications for use on APPLE II system. Such microcomputer systems and the associated use-developed applications can be further supported by participation in local computer use clubs that offer program exchanges. The largest computer exchange group in the country is the A.P.P.L.E., which is the Apple Puget Sound Program Library Exchange in the Seattle area. For about $25 per year, the 11,000 APPLE users who belong to the organization get access to software at a reduced rate, a subscription to the exchange's magazine and use of its telephone hotline.[3] These machines are likely candidates for individual surveillance systems for physician offices.

Table 9-1 illustrates data elements necessary for tracking epilepsy patients to monitor drug therapy blood levels. The data elements in Table 9-1 could be captured on a microcomputer using existing commercial software. A database management package could provide a surveillance program for this example.

How Can Data Elements Be Identified Effectively?

Effective use of microcomputer packages begins with identification of the essential data elements needed for the surveillance process. The following questions would assist the physician in identifying them.

(1) What is the information that is to be tracked?

(2) Where does it originate? It may originate from a prescription order or like blood pressure readings recorded by a nurse during a visit.

(3) What is the frequency of its occurrence? Will it be picked up at each visit? Could it be reported by phone? Should it be monitored weekly or monthly?

(4) What is the volume of the data broken down by data elements and number of characters? For example, a diagnosis may require 15 characters. A code used to represent a tracking item may require two characters.

(5) What is done with the information? For one case, this may be a step-by-step identification of a flow sheet development used to monitor a particular medical treatment.

(6) How long does it take to collect the information? Is it collected from an encounter for completion during the visit?

(7) Who collects it? Many items can be flagged by the physician and collected for a surveillance system by practice staff for data entry on a microcomputer.

(8) Are controls maintained? Is there a system of checking and/or validating items scheduled for surveillance?

(9) How is the information going to be used? Here, it is important to define some of the purposes of a surveillance system such as: providing patient notification; summary treatment results for the physician; and flagging for transfer to a referral site or consulting physician.

Answering these questions will assist the physician in identifying information requirements. Office staff should be able to provide many of these.

Once these questions are answered, a sequential list or itemization of the process for identifying, collecting, and retrieving the information can be prepared. Once these are prepared, specific resources in hardware and existing software packages could be reviewed.

Other chapters offer more specific detail in system selection. The important components are a clear identification of the information required and an organized explanation of how it is collected and subsequently needed. Technical means are then employed to support these components.

What Existing Resources Are Available?

As indicated, in the discussion of microcomputers, software packages currently exist to assist nonprogrammer users in employing the system.

Medical user groups for the APPLE and the TRS-80 computers have been established. MUMPS User Group (MUG) is another source for interested beginners. There is a Micro MUMPS newsletter which provides system and package information on the microcomputers and referrals to people who presently work with these applications. Other medical and computer groups exist and can usually be reached through vendors in the area.[4]

Literature searches on medical records and patient or disease surveillance systems will provide additional information and often the referred articles have excellent suggestions.[5]

Considerable software has been developed through federal government grants and is available for minimum distribution costs as well.

The technology today offers extensive practice management assistance at reasonable cost. More importantly, it enables more and better use of the patient data housed in the office practice.

References

1. Kuhn, IM, Wiederhold, G: "The evolution of ambulatory medical record systems in the U.S.," *Proceedings of the Fifth Annual Symposium on Computer Applications in Medical Care*, IEEE Computer Society, Washington, D.C., 1981.
2. Ashton, J; Brinkman, D; Balsam, J: "Choosing a medical office computer," *Interface Age*, September, 1981.
3. "To each his own computer," *Newsweek*, February 22, 1982.
4. Walters, R: "Goals of the medical software exchange forum," *Proceedings of the Fifth Annual Symposium on Computer Applications in Medical Care*, IEEE Computer Society, Washington, D.C., 1981.
5. Brandejs, JF; Pace, G: *Physicians' Primer on Computer—Private Practice*, Lexington Books, D.C. Heath & Company, Lexington, MA, 1979.

Chapter **10**

Quality Assessment and Quality Control of Patient Care

Elmer Gabrieli

Objectives

Discuss the usefulness of assessment of quality of care rendered in the office.

Need for Quality Control of Care

Medicine Lacks Empirical Feedback

The celebrated Viennese surgeons of the 1920s pioneered radical procedures such as gastrectomy and colectomy. This heroic period was promptly followed by postoperative morbidity/mortality reports as a numerical expression of outcome of surgery. Thus the goal changed. The "best" surgeon was the one with the lowest postoperative morbidity and mortality. Similar feedback for self-audit is still minimal in many fields of clinical medicine. For example, the author of this chapter, a pathologist, has made thousands of important diagnoses of leukemias and other malignant proliferations, with essentially no formal long-range feedback and with an unknown error rate. Some primary care physicians can follow up their cases but they also often doubt whether they have chosen the "best" therapy. There is no easy way for a practitioner to compare alternative case management.

Rapid progress in all areas of biomedicine during the last four decades has made it difficult for a single practitioner to develop personal experience and sharing experience at various conferences is also less and less adequate.

Computerization of the medical record is a quantum leap in all these problem areas. Electronic storage of clinical case histories opens new vistas since clinical experience has become readily retrievable and relative success rates can be determined. We can now easily examine our own performance and compare our own results with those of others. This is an exciting new possibility for self-improvement. Perhaps this is the long awaited electronic revolution in clinical medicine.

Data Entry in the Private Office

In the past, there was a major barrier between human language and the computer input. This barrier has been recently successfully eliminated,[1] and today, computers equipped with an automatic medical "translator" can accept direct natural language input. Building the first medical translator was a monumental task. However, it is now completed and clinicians can benefit from it. The function of this translator is to convert natural language text into machine-compatible symbols. The size of this translator is rather large, usually too large for an office microcomputer, but a bigger remote computer can carry out the translation for many office computers. This requires a distributed network of office microcomputers with a supporting large computer equipped with the translator. This design makes the network cost-effective.

In order to fully benefit from the office computer, first an electronic copy of the office record must be created. This electronic patient record includes the physician's notes, laboratory results, EKG findings, X-ray reports, consultations, hospital discharge summaries, and all other documentation pertaining to the care of the patient. A typist enters all the information from the paper "source document," letter-for-letter. The office computer is programmed to guide the input clerk by asking for the various types of data. On the screen of the computer "prompts" are displayed to guide the clerk. In our experience, the entry into the computer of all the office records generated and reports received during a typical busy day takes 1.5–2.5 hours of clerical time. (This may compare favorably with the traditional manual clerical tasks of filing all the reports received and constantly filing and refiling all the charts.) The result is a cost-effective electronic record system, ready for retrieval.

Retrieval

Computers can accept, store, process, and retrieve submitted data, but the true justification for computerization is data retrieval. Therefore, we should critically examine the benefits of machine-generated reports. In general terms, the automated reports for an office practice can be divided into three major categories:

(1) *Fiscal retrieval* such as automated patient billing, daily earnings, accounts receivable, or aging of accounts can be actually *derivatives* of the electronic medical record. All four data elements, i.e., the patient, insurance carrier, diagnoses, and services rendered are in the chart, readily recognized by the machine, and can be extracted for driving the appropriate billing programs. Fiscal retrieval alone should make computerization cost-effective.

(2) *Administrative/managerial retrieval* enables the busy practitioner to study his own operation. There is no limit to such administrative reports. Let us examine just a few simple examples:

(a) *Practice status report* shows the exact answer to an important question. This report tabulates the entire practice, by year (a "patient since ... "). This report makes it immediately apparent whether the overall size of the practice is growing, stationary, or decreasing. Certain practitioners or groups may even be interested in quantifying the different "losses" by requesting special reports such as all deaths, the number of patients relocating, or cases seeking continued help elsewhere.

(b) *Source report* is an organized presentation of the various sources of patient referral, including the number of cases per source, their diagnoses, and even subdivided by the types of payment category. This report should help guide the clinician in his commitments to various institutions and groups.

(c) *Geographic map report* shows the distribution of home addresses of all patients in the practice (Figure 10-1 A and B).

(d) *"Time between visits" report* is another use for management information. Usually, the instructions to the patient include the suggested date for the next visit. The choice of this date is guided by several factors such as the acuteness of the clinical condition, the risk of complications, or the chance of adverse drug reaction. Further, these decisions have an important economic aspect, both to the patient and to the practice. In individual cases, most of us recommend the date of the next visit somewhat intuitively. The computer can generate tabulations of revisit intervals by disease category or by some other factor such as patients with pacemakers or cases on a potent drug with known frequent adverse reactions. Careful review of the intervisit time report should change the physician's revisit instructions from an intuitive to a conscious, deliberate decision (Table 10-1).

This brief, sketchy listing of a few examples should indicate that the possibility of administrative-type reports is boundless, limited only by the imagination of the physician.

(3) *Clinical retrieval* is the most important benefit of computerization. This is the central theme of this chapter. The physician's electronic case history bank (the database) can be used for a wide spectrum of different reports, instantly, easily, at the office or at a remote site. We shall briefly review some of these reports.

(a) *Single patient's records* can be retrieved in several forms:

1. *Synopsis*, which contains the patient's name, and the list of current diagnoses and drugs (Table 10-2)
2. *Last Visit*, which shows the medical notes and the results of the diagnostic studies ordered during that visit (Table 10-3)
3. *Graphic display* of a "numeric variable" such as body weights, blood pressures, serum potassium values, or digoxin levels, showing all the numerical values during the last year (Table 10-4)
4. *Drug history* in chronic diseases tabulates all the drugs ordered in the past and later discontinued
5. *Full chart* is the retrieval of all the data on file.

The machine-generated synopsis has proven particularly useful when the physician has a home terminal. Emergency calls after office hours can be handled differently when the physician can use his terminal at home to retrieve the patient's diagnoses and current list of drugs by calling on the office computer.

Electronic case histories eliminate all the problems of traditional paper records such as lost laboratory reports, misfiled charts, the time wasted thumbing through a thick record, or the time spent on studying a chart kept by a partner which may not always be legible or explicit. However, it may take some minor adjustments until a physician feels comfortable in a computerized office. The transition may be made gradually. Initially, the paper charts may be retained, just for the sake of continuity, but the computer-printed synopsis can be inserted in to the paper chart and the machine can preprint the headings for the physician (such as "Complaints Since Last Visit," "Present Status," "Diagnostic Studies," "Change in Therapy," "Instructions," and "Next Visit"). This way, the physician can gradually become accustomed to the machine-generated records and abandon the paper record.

(b) *Statistical retrieval*: These reports are particularly valuable since they open up a new dimension; the ability to examine the entire practice. The diagnostic mix may be of interest, but the analysis of specific subsets is particularly interesting. In a cardiology practice, for example, the report of all cases of angina pectoris can also show the subset with coronary bypass surgery and compare them with those receiving conservative therapy. The

---

Figure 10-1 A Geographic distribution of patients of a medical practice. Computer-generated map of Buffalo shows the census tracts and the percentage of patients living in each tract.

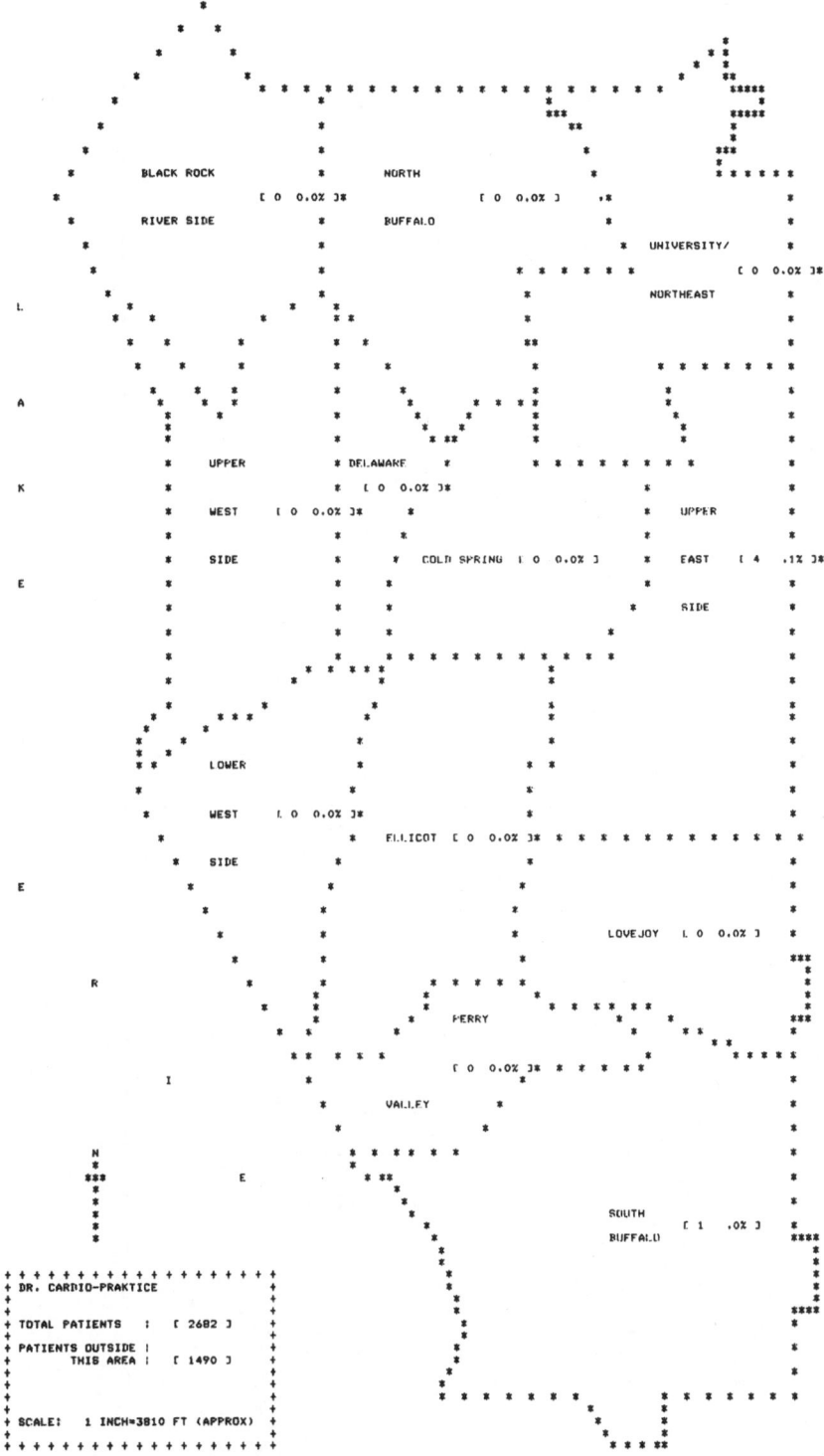

BLACK ROCK

[0 0.0%]

RIVER SIDE

NORTH

[0 0.0%]

BUFFALO

UNIVERSITY/

[0 0.0%]

NORTHEAST

UPPER

* DELAWARE

[0 0.0%]

WEST   [0 0.0%]

SIDE

COLD SPRING [0 0.0%]

UPPER

EAST   [4 .1%]

SIDE

LOWER

WEST   [0 0.0%]

ELLICOT [0 0.0%]

SIDE

LOVEJOY [0 0.0%]

PERRY

[0 0.0%]

VALLEY

SOUTH

BUFFALO   [1 .0%]

+ + + + + + + + + + + + + + + + + +
+ DR. CARDIO-PRAKTICE   +
+   +
+ TOTAL PATIENTS : [2682]   +
+   +
+ PATIENTS OUTSIDE :   +
+   THIS AREA : [1490]   +
+   +
+   +
+ SCALE: 1 INCH=3810 FT (APPROX)   +
+   +
+ + + + + + + + + + + + + + + + + +

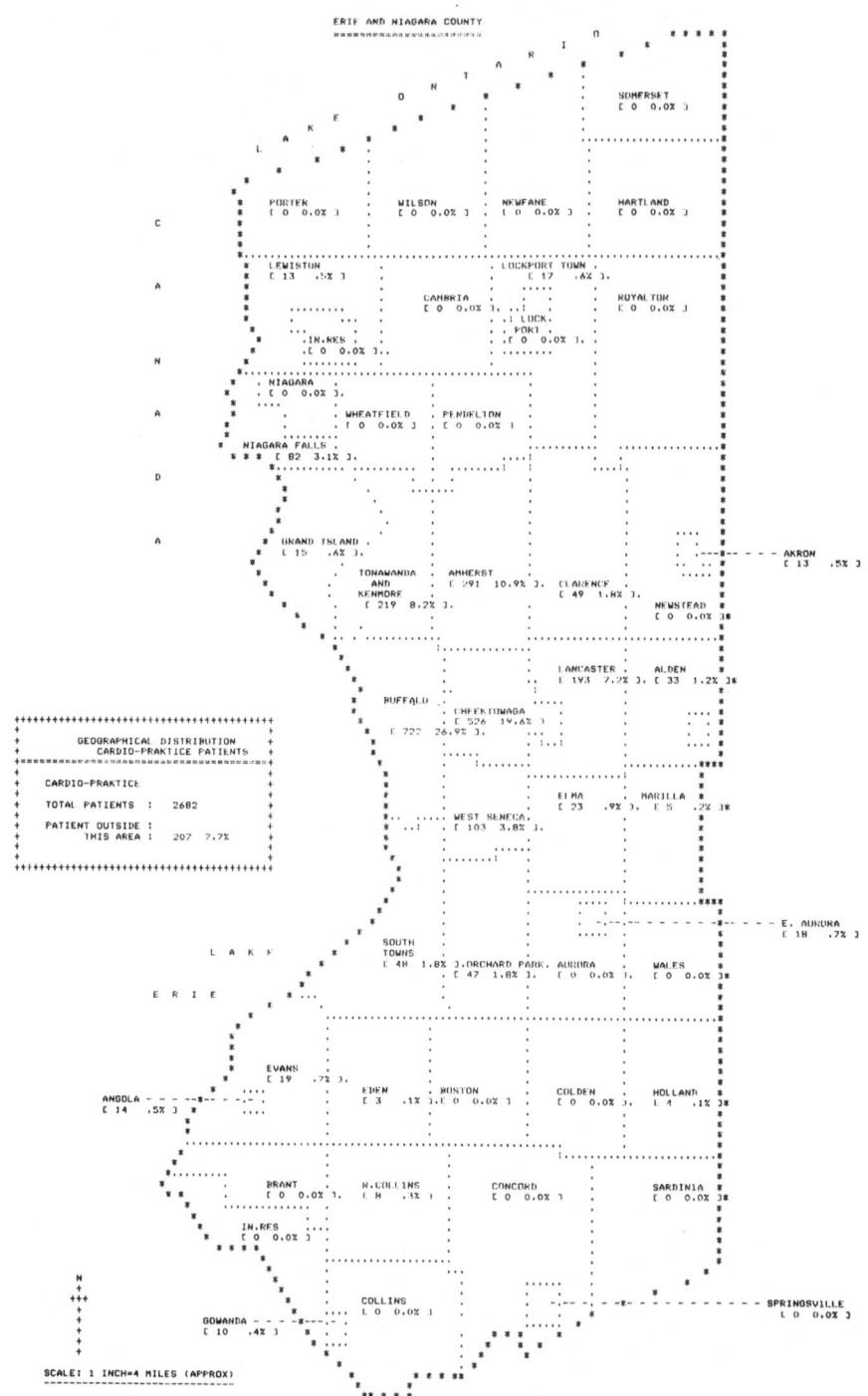

ERIE AND NIAGARA COUNTY

SOMERSET
[0 0.0%]

PORTER   WILSON   NEWFANE   HARTLAND
[0 0.0%]   [0 0.0%]   [0 0.0%]   [0 0.0%]

LEWISTON   LOCKPORT TOWN
[13 .5%]   [17 .6%]

                CAMBRIA   ROYALTON
                [0 0.0%]   [0 0.0%]
                            LOCK
                            PORT
IN.RES   [0 0.0%]
[0 0.0%]

NIAGARA
[0 0.0%]
            WHEATFIELD   PENDLETON
            [0 0.0%] [0 0.0%]
NIAGARA FALLS
[82 3.1%]

GRAND ISLAND
[15 .6%]   AKRON
    TONAWANDA   AMHERST   [13 .5%]
    AND   [291 10.9%]   CLARENCE
    KENMORE   [49 1.8%]
    [219 8.2%]   NEWSTEAD
                                          [0 0.0%]

                              LANCASTER   ALDEN
                              [193 7.2%], [33 1.2%]

        BUFFALO   CHEEKTOWAGA
                      [526 19.6%]
        [722 26.9%]

GEOGRAPHICAL DISTRIBUTION
CARDIO-PRAKTICE PATIENTS
                                          ELMA   MARILLA
CARDIO-PRAKTICE   [23 .9%], [5 .2%]
TOTAL PATIENTS : 2682
                              WEST SENECA
PATIENT OUTSIDE :   [103 3.8%]
    THIS AREA : 207 7.7%

                                                    E. AURORA
                                                    [18 .7%]
                    SOUTH
                    TOWNS
LAKE   [48 1.8%] ORCHARD PARK, AURORA   WALES
                               [47 1.8%], [0 0.0%], [0 0.0%]
ERIE

            EVANS
            [19 .7%]
ANGOLA   EDEN   BOSTON   COLDEN   HOLLAND
[14 .5%]   [3 .1%],[0 0.0%] [0 0.0%], [4 .1%]

            BRANT   N.COLLINS   CONCORD   SARDINIA
            [0 0.0%], [8 .3%] [0 0.0%] [0 0.0%]
IN.RES
[0 0.0%]
                COLLINS
                [0 0.0%]   SPRINGSVILLE
GOWANDA   [0 0.0%]
[10 .4%]

N

SCALE: 1 INCH=4 MILES (APPROX)

Table 10-1 Time Between Visits

```
VISIT INTERIM - FULL PRACTICE DR. 001                    1/24/1983

INTERIM              NUMBER
=======              ======

1 - 6 DAYS           8
1 - 2 WEEKS          7
2 - 3 WEEKS          20
3 - 4 WEEKS          12
1 - 2 MONTHS         146
2 - 3 MONTHS         135
3 - 4 MONTHS         159
4 - 5 MONTHS         106
5 - 6 MONTHS         89
6 - 7 MONTHS         149
7 - 8 MONTHS         83
8 - 9 MONTHS         77
9 - 10 MONTHS        70
10 - 11 MONTHS       61
11 - 12 MONTHS       82
OVER 1 YEAR          1254


AVERAGE INTERIM BETWEEN VISITS IS      427 DAYS
```

Note. This printout shows the intervals between office visits, with the number of patients in each category. Other printouts show these revisits by disease categories.

success of surgery became evident when we compared the drug therapy of the two groups, focusing on the need for vasodilators. The absence of angina, as one measure of success, became apparent. Comparison of the two groups by "return to work" was another aspect. This type of retrieval also provided the chance to examine more closely the subset of coronary bypass cases with continued or recurring angina. Perhaps the indication for surgery should be further sharpened. An automated surveillance of patients with pacemakers is also a useful project for the computer. Another example is the analysis of drug therapies. The machine can tabulate the list of "Changes of Drugs," along with the clinical reasons. The list of changes in drug therapy is an important tool for self-audit. Change of a drug may occur for several reasons, such as side effects or insufficient effectiveness. If this report shows that a particular drug had to be discontinued in several patients, the physician may compare that drug with alternative therapies.

The relative success of treatment is perhaps the most important self-audit. For example, we have generated a report for our cardiologists with regard to their own success rates maintaining their hypertensive patient within the

Figure 10-1 B Geographic distribution of patients of a medical practice. Computer-generated map of Erie and Niagara counties showing city and township limits and the percentage of patients residing in each community.

Table 10-2 Synopsis of a Patient Office Record

```
Pat#1391                                          Patient Synopsis
68-year-old                                    RETIRED STEEL WORKER

                           Dx Summary
                           ==========

   Visit    Diagnosis                        Status
-----------------------------------------------------------------------
 | 04/28/78 | CORONARY HEART DISEASE          |
-----------------------------------------------------------------------
 | 04/28/78 | HYPERTENSION                    |
-----------------------------------------------------------------------
 | 04/28/78 | APHASIA                         |
-----------------------------------------------------------------------
 | 04/28/78 | CEREBRAL ATHEROSCLEROSIS        |
-----------------------------------------------------------------------
 | 04/28/78 | DIABETES                        |
-----------------------------------------------------------------------
 | 09/12/79 | DIABETES                        | MILD IMPROVEMENT
-----------------------------------------------------------------------
 | 09/12/79 | MYOCARDIAL INFARCTION           |
-----------------------------------------------------------------------

                           Drugs
                           =====

   Visit    Drug Name      Dose         Frequency       Comments
-----------------------------------------------------------------------
 | 09/12/79 | DYAZIDE      |            | BID           |
-----------------------------------------------------------------------
 | 09/12/79 | DIGOXIN      | 0.25       |               |
-----------------------------------------------------------------------
 | 09/12/79 | DIABINESE    | 250        | QD            |
-----------------------------------------------------------------------
 | 08/18/80 | DIGOXIN      | 0.1        |               | REDUCE
-----------------------------------------------------------------------
```

Note. This printout shows two demographic data (age and occupation), the last seven visits, diagnoses, comments, and drugs ordered.

normotensive range (Table 10-5). These histograms of the last systolic pressures were informative. Upon reviewing them the cardiologists asked for a drug listing of those cases which had a systolic pressure of 170 or higher. The comparison of these histograms with the results of the other cardiologists was even more informative. Another interesting report was the long-range follow-up of the stress test cases.

When a drug was withdrawn from the market because of its side effects (e.g., Selacryn), the computer was able to notify all the patients on that drug to contact the office for an appointment.

Perhaps the most exciting report was the retrieval of high-risk cases prior to the influenza season. For this report, a set of criteria was chosen, such as chronic heart failure, advanced obstructive pulmonary disease, diabetes, or age over 70. Upon selecting these criteria for vaccination, the machine could identify these high-risk patients and a brief explanatory letter on the value of vaccination could be sent.

Co-occurrence of diagnoses is also an important statistical report. Co-occurrence of hypertension/diabetes/obesity is an example. Co-occurring

Table 10-3   Last Visit Report

```
07/13/1981 09:10 AM
Revisit                                    Pat# 1391


                         Vital Signs
                         ===========


Weight: 172                              BP: 130/90


                    History Since Last Visit
                    ========================

        FEELS MUCH BETTER- SPEECH IMPROVED


                    Studies Ordered/Results
                    =======================

  Study Name: CHLORIDE
             Result: 92 Low**

  Study Name: CHOLESTEROL
             Result: 329 High***

  Study Name: DIGOXIN
             Result: 1.0

  Study Name: GLUCOSE
             Result: 137 High**

  Study Name: POTASSIUM
             Result: 3.6

  Study Name: SODIUM
             Result: 151 High*

  Study Name: TRIGLYCERIDE
             Result: 148
```

Note. Last visit of the patient showing what the physician has
recorded (letter for letter) and the results of the laboratory
studies ordered; * following the laboratory result signifies
abnormal finding, ** signifies markedly abnormal finding.

diagnoses reveal deeper correlations among clinical manifestations (Table
10-6). These types of reports enable the clinician to benefit from the
statistical information.

The clinical reports become particularly helpful if they can be compared
with corresponding data from other clinicians. For example, the treatment
success of a physician in breast cancer, systemic lupus erythematosus, or
rheumatoid arthritis should be compared with corresponding reports of
others. This can be achieved fully anonymously, so that the individuals are
not recognizable. The author of this chapter is currently organizing a rational
database for clinical cardiology. The participants will be able to retrieve the
pooled clinical experience, a new dimension in medical education.

Table 10-4 Graphic Display of Digoxin Level

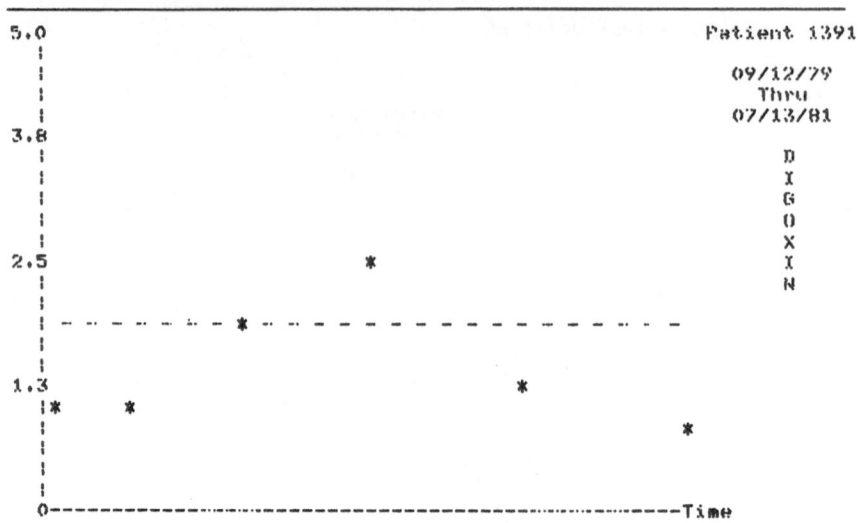

Note. Display of six serum digoxin levels assayed during 2 years. Each * represents a digoxin determination; the fourth value is outside the recommended therapeutic range (interrupted line).

Table 10-5 Computer-Generated Histogram of Hypertension

CUM %	LAST BP	
0.00%	240	
0.00%	230	
0.00%	220	
.31%	210	＊
1.24%	200	＊＊＊
2.17%	190	＊＊＊
8.07%	180	＊＊＊＊＊＊＊＊＊＊＊＊＊＊＊＊＊
15.22%	170	＊＊＊＊＊＊＊＊＊＊＊＊＊＊＊＊＊＊＊＊＊＊
21.43%	160	＊＊＊＊＊＊＊＊＊＊＊＊＊＊＊＊＊＊＊
32.92%	150	＊＊＊＊＊＊＊＊＊＊＊＊＊＊＊＊＊＊＊＊＊＊＊＊＊＊＊＊＊＊＊＊
51.86%	140	＊＊
73.60%	130	＊＊＊
86.65%	120	＊＊＊
95.65%	110	＊＊＊＊＊＊＊＊＊＊＊＊＊＊＊＊＊＊＊＊＊＊＊＊＊＊＊＊＊＊
100.00%	100	＊＊＊＊＊＊＊＊＊＊＊＊＊＊

PATIENTS

Note. Histogram of last systolic blood pressure measurements in a physician's office practice; each * represents one patient. Cumulative percentage is shown in first column.

Table 10-6 Co-occurrence of Paired Diagnoses for Dr. 001

	Number	Percent
Hypertension	688	100.00
Status post bypass surgery	120	17.44
Chronic ischemic heart disease	133	16.43
Disorders of lipid metabolism	88	12.79
Diabetes mellitus	77	11.19
Obesity	55	7.99
Old myocardial infarct	53	7.70
Conduction disorder	44	6.39
Heart failure	41	5.96
Angina pectoris	28	4.07
Chronic rheumatic heart disease	16	2.33
Chronic endocarditis	15	2.18
Atherosclerosis	15	2.18
Cerebrovascular disease	9	1.31
Acute endocarditis	5	.73
Arterial embolism	5	.73
Cardiomyopathy	4	.58

Note. Computer-generated listing of additional diagnoses in 688 hypertensive patients, ranked by frequency.

Clinical use of computers is still in a very early stage. We can anticipate, however, that during the 1980s, physicians will fully accept the technology and exploit its potential. The office computer, linked to a large support machine, will assist the fiscal and managerial tasks, and it will enable the clinician to constantly analyze his own performance, to constantly compare his own success rate with others, and to learn from his own errors in judgment. Further, the office computer will provide the newest information on any diagnostic or therapeutic question, by interrogating the electronic knowledge bank via the office terminal. Information-dependent clinical decisions will be supported by the office computer, making the pertinent information instantly available: the physician's own experience, the experience of others in similar cases, and the newest published information on that particular problem.

The computer-supported clinician will optimize his clinical decisions, but clinical judgment will remain entirely the task of the physician.

Reference

1. Gabrieli, ER: "Praktice—an advanced clinical information system," in *Computers for Medical Office and Patient Management*, edited by SB Day and JF Brandejs, Van Nostrand Reinhold Company, New York, 1982.

Chapter 11

Patient/Parent/Family Educational Assists

Byron B. Oberst

Objectives

Present a methodology to document medical treatment in educational programs and to measure the functional outcomes of these programs.

Introduction

With the great awareness of "health risk factors," popularity of "healthing-keep well programs," and the need for patients to avoid health destructive habits, modern medicine of today, and more so in the future, will be utilizing health care delivery. The patient and his or her family are much more informed than ever before. They are more sophisticated regarding health matters. The patient and his or her immediate and extended family expect the physician to provide various types of educational approaches and information to his or her own specific health needs.

Health education needs to be provided on a general prevention basis as well as on a specific disease state need. Many approaches may be utilized including regular programmed health instructions on avoidance of health problem areas, proper self-examination, understanding of one's own disease process, and similar phenomena.

The physician needs assistance in managing these educational approaches, in keeping track of the variety of materials and methodologies used, and in providing for some type of longitudinal outcome determination. The computer can provide the means of monitoring and logging these various educational usages.

Educational Assists

The computer can assist the physician in a number of ways:

(1) By printing out instructional materials for health supervision visits such as guidelines for toilet training, discipline techniques such as behavior modification, and school readiness factors.
(2) By indicating needed educational materials for certain medical problem areas such as patient understanding, hypertension, diabetes, or arthritis.
(3) By providing methods for logging and following the influence of the various educational programs or materials utilized.
(4) By keeping track of the apparent results and documenting the effectiveness of the medical treatment programs including the educational aspects.
(5) By documenting the patient's functional outcome based upon identified coordinates, the patient treatment outcome determination will become a need of the future. In this manner, patient compliance, cost factors, and quality control can be evaluated. *Example:* In my office, infants, children, and youth are followed by eight functional outcome areas: five regarding patient related areas, and three regarding the patient's family areas.

Functional Outcome Factors

Patient-Related Areas

1. Physical Aspect
   1.1 Excellent
   1.2 No limitations
   1.3 Mild handicap
   1.4 Severe handicap
   1.5 Chair
   1.6 Bed
   1.7 Total incapacity
2. Intellect Aspects
   2.1 Gifted
   2.2 Above average
   2.3 Average

    2.4 Below average
    2.5 Moderate impairment
    2.6 Severe retardation
    2.7 Total incapacity
3. Emotional Aspects
    3.1 Total independence
    3.2 Mild support needs
    3.3 Modern support—some counseling
    3.4 Frequent counseling
    3.5 Hospital shelter environment
    3.6 Total incapacity—custodial care
4. Educational Aspects
    4.1 No needs
    4.2 Regular class help
    4.3 Resource room/teacher
    4.4 Engineered room
    4.5 Special education programs
    4.6 Educational handicapped programs
    4.7 Other
5. Cultural/Social Aspects
    5.1 Total independence
    5.2 Moderate supervision needed
    5.3 Structure needed
    5.4 Foster care shelter
    5.5 Foster caree
    5.6 Institutional care

Family-Related Areas

6. Economic Means
    6.1 Excellent
    6.2 Moderate
    6.3 Questionable
    6.4 Substandard
    6.5 Government aid welfare
7. Family Resource Attitude
    7.1 Excellent
    7.2 Good
    7.3 Average
    7.4 Questionable
    7.5 Passively hostile
    7.6 Actively hostile
    7.7 Hostile
8. Family Stability
    8.1 Excellent

8.2 Good
8.3 Average
8.4 Questionable
8.5 Poor

By utilizing a voucher-type input device, these Functional Assessments can be logged into the computer and changed as the patient responds to the treatment program. The following example demonstrates this capability.

Functional Assessments—Infant Age 1:

Patient-Related Areas:

Physical function—no limitation
Intellectual category—average
Emotional category—mild support needed/due to age
Social/cultural—moderate supervision needed/due to age

Family Related Areas:

Finances—moderate
Family attitudes—good
Family stability—good

[Note: These Functional Assessments can be shifted as the treatment program procedures outcome improves and changes. With this capability, a patient's status related to an intermediate or long-term medical problem can be recorded before and after a treatment program.]

Part IV

Introduction to Planning, Vendors, and Implementation

Chapters 1 through 12 helped to highlight the usefulness of computers in private practice in the accounting, administration, and health care delivery aspects. The prospective physician considering computerization of his or her practice should spend considerable time and creative thought upon the foregoing chapters.

Chapters 13 through 16 provide the How-To-Do-It approaches. Any major step, such as computerization of an office practice, should include a careful need analysis and detailed plans. Part IV outlines the remaining aspects to be accomplished.

Chapter 12

Planning for Automation: The Total Office Practice System

John M. Long

Objectives

Provide suggestions on how to put your wishes into a useful plan.

Put automation and the computer into perspective in relation to your office systems, and discuss other major components of the system, including personnel training and space and service needs.

Introduction

Personal Involvement in Planning

Now, after reviewing the material in this book, you will want to pull together your thinking into a plan. The plan can be very simple or quite complex, depending on how extensive your desire is to get involved with automation. The extent of your personal involvement in planning will also depend on the extent of your automation plans. If you are primarily concerned with automation of accounting and administrative functions, as discussed in Chapters 3, 4, and 13, a practice consultant, with assistance from your staff, can do much of the work of planning for you once you have made the basic decisions. If you wish to go further into automation of those modules

(Chapters 5 through 11) directly related to your practice, your involvement will be more extensive and eventually you and your staff will almost certainly also need the help of an outside consultant.

I suggest that you use an iterative approach to planning as discussed below and that you (especially you) approach planning from a global or total office systems approach.

Why Do I Need a Plan?

In the first chapter we discussed the need to focus on the system to be automated, more specifically, to focus on the function(s) of the system under consideration rather than on one of the tools (computers) used to accomplish the function. Some of the examples of computer applications in physicians' offices found in the literature seem to be essentially trivial from a clinical practice point of view, e.g., keeping a catalog of a medical reference file on a computer. Such applications were probably done for the fun of it. Someone had a computer and was looking for a way to use it; a solution looking for a problem.

Such uses are all right if the motivation was to test or experiment with a computer or simply to play with the computer. For serious use, however, this approach is "putting the cart before the horse." One should first focus on his or her office procedures, identify them, and decide which ones need to be improved and which ones might benefit from automation. Those selected for further study should be carefully analyzed concerning all aspects including space, equipment, personnel, skills, policies, and procedures. At all times keep in mind the interaction of the system under study with other office systems. Special attention must be given to the impact that changes in the system under study will have on the other systems in your office. I call this focus on all aspects of the system, not just those susceptible to automation, and on the interaction of each system with the other systems, the *total system approach*. In order to accomplish this, planning is required.

Planning Phases

When I take on a new project, I usually must go through several iterations before I achieve an acceptable result. The same is true for planning.

Phase I: Set Objectives

The first phase of your planning probably will involve you alone. Already, as you have read the material in the book, you have been formulating your ideas of what, if anything, you will want to do. You are probably already using a computer service to do your accounting and you may not want to disrupt it. Your plan may be to purchase a personal computer and essentially play with it to become more familiar with computer concepts. Incidentally, this step is not necessary. You may be thinking of modifying your accounting system so that you can expand into automation of the medical record and other clinical practice applications. You may be planning a new office and decide that you want to go with a paperless one. A brave decision! Whatever you want to do, it is important for you to think through in your own mind what you want to do. Only you can do this. All of the things that follow will depend on these decisions. Others can help to refine and develop the plan but you must make the first basic plan. Some important things to consider as you formulate this first basic plan are: What are my motivations? What do I want to accomplish? How much time can I give to the project? How much money can or will I commit to the project? How will my staff accept the changes? What can I do to help my staff accept the change or am I willing to replace staff if necessary in order to achieve my goals? How will my patients react? Should the system be transparent to patients? Do I have any spare space for equipment?

Phase II: Gather Data About Your System

Once you have a good idea of what you want to do, you can move to the second phase or iteration of your planning. At this point you will probably need a consultant to help you to prepare a written plan. He can help you gather the data you need to determine exactly what is involved in accomplishing the objective you have set. The data gathering involves information about your practice as well as information about what is available to help you accomplish your objectives. Most of the planning suggestions listed below are based on previously discussed material so that the list also serves as a partial summary.

Some of the modules of your office system which you will want to consider for partial automation are listed in Table 12-1. This is a good list to consider since all modules in the list have actually been automated (i.e., there is no "blue sky" here).

Table 12-1   Major Components of a Physician's Office System

On Line Accounting (Chapter 3)
Administrative Application (Office practice management) (Chapter 4)
Clinical Application (Patient Care) (Chapter 5)
Medical Records (Chapter 7)
History Taking (Chapter 8)
Supervision/Tracking Patient Care (Chapter 9)
Administrative Needs (Chapter 4)
Office Practice Management (Chapter 12)
Educational Needs and Professional Development (Chapter 6)
Patient/Parent/Family Education (Chapter 11)

Table 12-2   The Anatomy of a System

---

(1) Policies and procedures
(2) Description of flow and volume of patient through system (if relevant)
(3) Description of flow and volume of material (e.g., specimens) through system (if relevant)
(4) Description of flow and volume of data through system
(5) Means of entry into system
(6) Space and equipment needs (include equipment maintenance)
(7) Staff requirement (level of training required and retraining needs)
(8) Provisions of service—Protocols
(9) Recording relevant data
(10) Storage and retrieval of data
(11) Means of tracking both costs and charges
(12) Interaction with other office systems

There may be other systems or modules that you can think of, such as registration and appointments. These can be thought of as separate modules if you like, or as a part of one of those in the list.

You must examine the current status of each of the components of the system in detail. Table 12-2 is repeated from Chapter 1. You may want to look back at the section entitled, "Essential Elements of a Physician's Office System," because a part of the planning process involves the anatomical study of each system (or module).

Phase III: Gather Data About What Is Available

The third phase involves finding out what is available. This book provides you with data on what is possible and, to some extent, what is available now to help you. Oberst provides an in-depth case study of what one pediatric practice has done. Hoskins has discovered about 150 companies that provide computer services to private practice physicians. (No doubt almost all of these provide financial services only.)

You or your consultant need to see what is available to you in your location and in your price range.

Phase IV: Develop Your Plan

As you develop your plan keep in mind the cautions presented in Chapter 1. Other considerations you will want to look at are:

(1) *Staff/Patient Considerations.* There is a myriad of examples of attempts to automate a system without carefully considering the human interface. All of us have at one time or another enjoyed the frustrations of computers gone awry. The man–machine interface was not adequately designed. There is a common human characteristic which resists change no

matter how good it is. This is especially true if the person who must change was not involved with the decision to make the change or if the person is afraid. Your staff may resist automation because they are afraid it will replace them or make them obsolete. They worry that they are not capable of learning how to do things the new way. These perceptions are not trivial problems. Be sure to involve your staff as early and as much as possible. Give them time to adjust their thinking. Don't try to do too much at once.

(2) *Training Requirement.* Your staff will need to learn to do things a different way if you automate parts of your office system. If the system itself is clearly defined, it becomes a relatively easy task to *define* the new skills to be learned and the new tasks to be done. The learning itself may not be so easy and it certainly comes at some cost in staff time.

(3) *Space and Alterations.* New equipment takes up space. Computer terminals use more space than some crowded offices have. In addition, cables may be required, as well as additional wiring, and new telephone lines, and, in some unusual cases, additional air conditioning may be needed.

(4) *Outside Services.* If you get your own computer, you will need maintenance contracts for the equipment and programming help on occasion. Computer supplies must also be provided. If you purchase computer services and install terminals, you need arrangements for maintaining the terminal, and for outside help when problems occur.

(5) *Backup System.* Something many people do not think about is system redundancy for backup of the system. When systems are operated manually, the backup system may consist of calling in a former employee to help out or fill in or some other similar casual arrangement.

Automated systems backup must be provided in a more organized manner. It is necessary to have some system redundancy. It may consist of an extra terminal in the office in case one doesn't work. It may consist of reverting to a manual system temporarily. (Incidentally, this need is another reason for having good manual systems before you automate.) One common way for providing backup is the duplication of all files periodically (daily or weekly) with interim transactions logged and saved until the next backup is done. If manual backup is used, the files are printed out and these paper files are used during the computer down time. One of the most difficult problems to deal with when a computer goes down is restoring the system after it is back in operation. A system must exist which will restore the computer to its status at the last backup and then enter all transactions which occurred from then up to the present. If you use a computer service, instead of your own computer, the service company worries about all of these things and this is one good reason to use the service.

The cautions listed above, in Chapter 1, and throughout the book are mentioned not as a discouragement, but as a caution. Nothing spoils an otherwise successful project like a discovery too late that you must spend more dollars on costs not initially anticipated.

The Plan

The product of this planning process will vary. If you plan to purchase or lease a computer and software, or if you plan to purchase computer services, you will need to provide the vendor with the following information. Therefore, this list provides the minimum content of your plan.

(1) A description of your current system even if it is a rudimentary manual operation
(2) A description of all computer-related equipment currently used
(3) A description of your future systems. Where you would like to wind up and in what time frame?
(4) Estimates of current and future work loads in as much detail as possible
(5) A statement of minimum acceptable performance requirements

More details are needed for more elaborate systems. Your personal involvement is needed in those aspects where basic decisions are made and in those phases which directly impact your clinical practice.

Chapter 13

Be Prepared to Give, Not Receive, the Sales Pitch

John M. Long

Objectives

Examine what others have done. Look at some points related to vendor selection. Outline the steps to follow in getting started. Provide an overall summary of the book.

Introduction

What Others Have Done

Perhaps it would be helpful at this point to review what others are doing. They have taken a number of different approaches. In Appendix B, a case study for a pediatric practice is described. Here we will give a more general review of what others are doing. The appropriateness of any one approach is directly related to the goals each individual wishes to achieve, based on his or her individual circumstances, of course.

Purchase a Computer System

There are a number of reported cases where the physician has purchased his own computer. Many of these are small personal computers such as the

Apple series, TRS-80, or similar models by other manufacturers. Often the motivation is based on personal interest, and the use quite limited. Typically, a physician purchases a small computer and experiments with it by keeping a reference file on it, or by storing and processing data related to a special interest area of his practice. The out-of-pocket expenses are in the range of a few thousand dollars.

There are more sophisticated examples as well. Some in-house stand-alone computers are quite expensive and sophisticated. Studney[1] reports an operational "paperless" physician's office in British Columbia. The computer with lines, terminals, and office modifications cost about $300,000. Staff training, added utilities expenses (additional air conditioning), and other expenses about equalled the equipment costs. This case represents the high extreme in costs, but for this expenditure the office operated without the need for any paper records except for a patient encounter form completed by the physician. Patient data (demographic, history, medical, treatment, and financial) are all kept in computer files. Billing, scheduling, and monitoring are integral parts of the system.

Purchase Computer Services

The method most commonly used today is to purchase computer services. This can be done in one of the following ways.

(1) Purchase of an off-line remote computer service; that is, no equipment is located in your office. This system has been commonly used for a number of years. The office staff prepares data on standard forms which are picked up by a courier service and taken to a central point for computer processing.

(2) Tie into a remote computer service by means of an office terminal. This is becoming more common. In some states the Blue Cross/Blue Shield group is encouraging this approach. (It has been reported in one southern state that the Blues are coercing the doctors to install terminals in their offices at their own expense. You do their work for them and pay to do it!) The advantage of this approach over the first one is that your office staff can enter data directly into the computer, thereby reducing errors since the office staff can check and validate the entry. It is also possible, depending on the arrangements, to interrogate the computer database from your office terminal for needed data. Most computer services available to physicians today are limited to billing and other closely related administrative functions.

(3) When a service bureau is used, billing, etc., may be the only practical application you can make of the system. Oberst[2] reports a very creative use of office terminals tied into his hospital's computer. The hospital has developed a wide variety of applications software for its own use. He has found that the cost of modifying the software for office use is only a small

fraction of the cost of the original development. In this way, he has been able to operate a wide variety of applications beyond billing. These are more extensively reported elsewhere in this text.

(4) Another variation of the shared computer approach is to form a computer cooperative with other physicians. A few of these have been reported. The Primary Care Cooperative Information Project[3] in the upper New England area is sponsored by the University of Vermont. Its start-up funding came from the federal government. The initial experiment with automation involved the purchase of accounting services from a Boston service bureau catering to physicians. They are currently attempting to develop their own resources. The Rhode Island Health Services Research, Inc.[4] reports on its efforts to develop a computer cooperative. I have heard of a group in California, and there are others attempting to do this. Your own professional organizations are the most logical ones to develop such systems. This trend is most promising and should be carefully watched.

Your Should Share Software—COSTAR

Except in unusual cases, if at all, it does not seem practical for an individual physician's office to attempt to develop its own software. As you will note, almost all of the examples given use software already developed. Some modifications to satisfy your individual requirements are often needed but these must be kept to a minimum. The only possible exception that I can think of would be in an experimental or developmental mode such as the paperless office in British Columbia previously cited.

COSTAR (*CO*mputer *ST*ored *A*mbulatory *R*ecord[5]) is a standardized package of computer software designed to provide comprehensive and integrated information processing support for the medical, financial, and administrative needs of an ambulatory practice. It has the following modules:

Registration
Scheduling
Medical Records
Accounts Receivable/Billing
Report Generator
Systems Maintenance (allows maintenance)
Medical Query Language

At this time, 18 physician offices are using at least one module of COSTAR. Typically, these offices serve 5000 patients with 20,000 encounters per year.[6]

Another variation of sharing software is to develop mechanisms whereby physicians can share their software with each other. I know of two efforts at

present: Apple Medical Users Group Software Exchange[7] and the Medical
Software Exchange Forum.[8]

Sharing Information from Medical Data Banks

Still another variation of sharing related to physician office automation is the
development of large data banks of medical information available to a private
practice physician, provided he has the means (usually a terminal, phone
coupler, and a phone) to interrogate the data bank.

Certainly the most ambitious relevant project today is the announcement
by the American Medical Association and General Telephone and Elec-
tronics of their joint venture project in this area. Others are now marketing
data regarding toxic materials, occupational hazards, and similar topics
which can be accessed from an office computer terminal.

Many consider the establishment of large automated data banks of all
kinds available via phone and a terminal to be the hottest current area of
automation. News, stock quotations, current tax regulations, as well as
medical data are all either on the market or in an experimental stage. The
principal drawback to their full exploitation is the potential disruption of
existing systems of communicating such information. The AMA decision
may well be the most significant one it has made in many, many years. It
seems obvious at this point that physicians will eventually want to either
have a terminal in their offices or have access to one. This is true even if you
purchase your computer, because we will probably want to access the data in
the information banks. The need to eventually have a terminal in your office
to access medical data banks, to communicate with insurance carriers, and
the like may influence your decision as to whether or not you should get your
own computer or purchase the service.

What to Do

In the previous sections we have attempted to provide you with the necessary
background. Now you are ready to proceed. The following sequence would
be a reasonable one.

(1) Review your motivation at this point. Do you want to proceed? How
much time do you have to devote to the project?

(a) If you want to learn more, you may want to purchase a personal
computer and experiment with its use or you may wish to attend a
seminar or conference devoted to automation in the physician's office.
(b) If you are primarily concerned with billing and accounting functions, you
may wish to purchase a computer service from your accountant or some
other source. Hoskins (personal communication) has identified about

150 companies which offer these services to physicians. Since automation of this area has little direct impact on the way to practice medicine, others can do almost all of the required work for you. If you elect this option, be sure to reread Chapters 3, and 4, as well as review the caveats in Chapter 1 and the comments regarding the use of consultants under Item 7, below.

(c) If you wish to consider some of the applications beyond billing, you will need to make a much larger personal time commitment even with the use of a consultant. This is necessary because you must first understand your current systems in the total systems sense (see Chapters 1 and 12) and because you will not want to delegate the decisions regarding how to automate systems which involve the way you practice medicine.

If you selected Motivation a, you need go no further in this sequence at this time. If, however, you selected Motivation b, you can skip to Item 6 below. If you selected Motivation c, you have a bit of homework to do at this point. Continue to Item 2.

(2) Formulate a general plan, that is, Phase I of the planning process described in Chapter 12. Decide what you want to accomplish, which systems you wish to automate. Select a time frame which is realistic for you in terms of the time you and your staff can devote to the project. Be sure to consider the disruptions which changes may make for you, your staff, and your patients. You need to decide the sequence you wish to follow, which systems need to be improved, which are more easily done, and what is the logical sequence to follow. All of this will help you in selecting the sequence. Review of Chapters 3 through 12 will also be helpful.

The plan need not be elaborate or very formal at this stage. I highly recommend that anyone who wishes to go this far with automation hire a good consultant. He can help you to formalize the plan. What you need to do at this stage is to think things through so that you can tell the consultant what you want to do. Chapter 12 discusses planning in more detail.

(3) If you are thinking of any automation beyond billing, you should carefully consider using COSTAR[5] as a software package. It isn't the final word and you may reject it, but you cannot afford to ignore it.

(4) Carefully review a real live working system. Oberst describes a very interesting approach which is *working* for him. His system is based on some unique circumstances which may not allow you to copy it. It will open up your thinking to the possibilities, however. If you want to consider going the "paperless" office route, get in touch with the British Columbia clinic[1] and review that system.

(5) You should also look into possible participation in a computer cooperative. Although the concept is an excellent one, the present level of development of the idea seems quite primitive for the most part. Also, you may have enough of a pioneering and independent spirit to want to have more control than a cooperative will allow. Eventually, a network of shared computer hardware serving private practice seems inevitable.

(6) It would be a very good idea to consider a system which will allow you to access banks of medical information, as well as to communicate with the computers at insurance companies and with your local hospital computers. This is not a very complex problem and may entail nothing more than a terminal costing as little as $1500 plus a phone coupler and, possibly, a data phone. Such capability seems essential if we are to keep from being overcome with paperwork.

(7) Hire a consultant. Acquisition of a computer system or a contract with a service is a commitment of significant sums of money over a long period of time. The criteria for selection are technical and there will be many vendors vying for your business. A good consultant can save you enormous amounts of time and will probably save money. The converse is also true for a poor consultant. The best criteria, in fact the only criteria, for evaluation of a consultant are client references, preferably from physicians with similar practices.

Everybody has a friend or relative in the business. Make an early decision on how much weight this will carry in your evaluation. If it is a significant factor and the friend is reasonably competent, you may save time by deciding to go with him now. It is possible that you could wind up with a good custom-designed computer system by working closely with a friend. It is also possible that you will lose both the friend and your money and have to start over.

One last caveat regarding how you select a consultant. Those you normally rely on for advice in such matters may not be able to give you the best advice. Your accountant's answer will probably be biased, depending on whether he does or does not provide computer services. Also, an accountant probably cannot advise regarding automation of nonaccounting functions.

(8) Establish internal systems responsibility. By the time you have reached this stage, you should have sufficient knowledge to properly direct the automation process. Since you probably cannot get involved with the intimate details, it is good to have some staff person (the clinic administrator if you have one, or his or her equivalent) who is committed and will be responsible for the successful operation of the system. This person should have some computer training if possible and should accept the responsibility willingly. He or she should also be familiar with the system applications specified in your plan and be involved with the system from the definition phase.

(9) With the help of the consultant (if you use one) and the staff person serving as internal project coordinator (if you use one), the plan should be formalized in the sense that what you want to do is written down. Be sure to involve your staff in this process in order to keep them from being threatened by the plan and to ensure their cooperation.

The plan should state what you want to do, breaking it down into the component steps. This might be a listing of the modules to be automated, the order in which you plan to proceed, and an estimated timetable for its accomplishment. You may need to include steps prior to the automation which allow you to improve or document existing manual systems in preparation for the automation.

(10) Size the system. If you do not have a good idea of the size of your files, the number of pages of printing, and the amount of data to be entered, you are not ready to talk to vendors. In fact, it could be dangerous to your financial health. Gamesmanship is still a factor in business and the seemingly naive computer user has an automatic disadvantage. This is one area where a consultant will be very helpful. You cannot evaluate a supplier or a hardware component with any hope of success unless you fully define and understand your system needs and use these criteria for your evaluation.

(11) Decide on your approach as to whether you wish to acquire a computer system or select a computer service.

(a) Acquire a computer system, either by a lease or purchase, with an existing package of appropriate software, i.e., existing programs to produce the results you want. Our bias is that getting your own computer is not suitable for very many private physicians' offices. If you go for the paperless office, perhaps it is appropriate. The decision as to whether you purchase or lease the system is essentially an economic decision involving tax implications and can be analyzed for you by your accountant. Keep in mind that computer hardware tends to become obsolete over a 4–6 year period. If the obsolete system works for you, you may not care about obsolescence. Also, even if you plan to replace it in a few years, it may still be more economical to purchase!

(b) Acquire a computer service from a medically oriented service bureau, from your hospital, or from a medical computer cooperative. This decision depends on your plan. No matter which is selected, you are buying the use of a combination of hardware and software capable of performing the jobs defined in your plans. The primary question in the evaluation of the system or service is "can the system or service do the job in the time frame that I require?" The best answer is obtained from a group of satisfied customers with identical or similar requirements. However, there is always something that makes each practice unique so that this "best" answer is not always evident. Similarly, almost all computer services are unique. In spite of the attempts of computer professionals to define and standardize, there are no universal criteria or methods for the evaluation of computer systems and services. There are only general guidelines.

(12) Prepare a proposal. This puts you in the driver's seat! You tell the vendor what you want to do. To obtain a rational response from a set of vendors, you must tell them what you need. The most common approach is to prepare a request for proposal. This document should contain the following:

(a) A description of your current system even if it is a rudimentary manual operation.

(b) A description of your future systems stating where would you like to wind up and in what time frame.
(c) A description of all computer-related equipment currently used.
(d) Estimates of current and future work loads in as much detail as possible.
(e) A statement of minimum acceptable performance requirements.

The request for a proposal will provide the information required for a vendor to make an intelligent response. Its preparation will require the buyer to gather enough detailed data to make an intelligent evaluation of vendor responses.

When these steps are done you are ready to select a vendor and proceed with your project. The above steps are not hard and fast. As stated in Chapter 1, very little is already established with regard to clinic automation. It follows that the steps are not that well established either. Finally, this material is designed to allow you to *proceed* with knowledge and understanding. It is not designed to discourage you.

Acknowledgment

The author is indebted to Joseph R. Brashear, Seven Pools Information System, Inc. for some of the material presented in this chapter.

References

1. Studney, DR: "The paperless physician's office," presented to *the Fifth Annual Symposium on Computer Applications in Medical Care*, IEEE Computer Society, Washington, D.C., 1981.
2. Oberst, BB: "Computer application to private practice," presented to *The Standards Committee, Annual Meeting Society for Computer Medicine*, 1981.
3. Bise, B; Nelson, E; Kirk, J; Chapman, R; Hale, F; Stamps, P; Wasson, J: "The primary care cooperative information project: A model for service and research in primary care," pp. 61–63, *Proceedings of the Fifth Annual Symposium on Computer Applications in Medical Care*, IEEE Computer Society, Washington, D.C., 1981.
4. Manire, L; Colt, A: "A small computer cooperative for health agencies," pp. 64–66, *proceedings of the Fifth Annual Symposium on Computer Applications in Medical Care*, IEEE Computer Society, Washington, D.C., 1981.
5. Kerlin, B; Greene, P: "*costar: An Overview and Annotated Bibliography*," The MITRE Corporation, November 1980.
6. Fiddleman, RH: "Who uses COSTAR and why." *Proceedings of the Fifth Annual Symposium on Computer Applications in Medical Care*, IEEE Computer Society, Washington, D.C., 1981.

7. Stoneburner, L: "Apple Medical Users Group: Software exchange," p. 76, *Proceedings of the Fifth Annual Symposium on Computer Applications in Medical Care*, IEEE Computer Society, Washington, D.C., 1981.

8. The Medical Software Exchange Forum—Organizing Council, p. 76, *Proceedings of the Fifth Annual Symposium on Computer Applications in Medical Care*, IEEE Computer Society, Washington, D.C., 1981.

Chapter 14

Problems with System Implementation

Robert A. Reid

Objectives

Examine potential problem areas and implementation to endeavor to prevent problems rather than cure them.

Introduction

Once you have decided to install a computer and have selected your system, you will be faced with your biggest problem: making it work in your practice. Heretofore, you have been facing problems in theory and problems in finance: Does the system have the capabilities you will need? Can you afford it?

Now you will be dealing with problems of fact. You must somehow get your new system going. Even though the *system* works, the *implemented system* may not work because of inaccurate data. System implementation may disrupt your practice and threaten you financially. Your key employees may get increasingly unhappy. They may (will?) quit, or worse, they may stay around and spill a cola soda onto your $5,000 disk drive. Once the system is in you may slowly realize that it was all a mistake. The system may be the wrong size or too inflexible.

The purpose of this chapter is to let you consider these potential problems before they arise so that you can plan ahead and anticipate them when they do.

We should say at the outset that these problems need not be severe, even if they occur. A well-planned system implementation can be very smooth. Each of the problems in this chapter usually do appear, but they can be handled simply if they are anticipated and recognized early. This chapter is directed toward prevention, *not* cure. With data systems as in health care, a failure of prevention can be costly.

Avoiding Problems

There is a general principle which you may follow to avoid problems with implementation: *visit a site where the system is already installed.* This visit is made to help you with the process of implementation and should be made *in addition* to any visits necessary to select the system.

On this visit, find the person responsible for selecting the system. Ask about the problems associated with system installation. In particular ask:

(1) How long was the interval between system delivery and full implementation?
(2) Were there unexpected costs involved in implementation?
(3) Did implementation disrupt the practice?
(4) Did any employees quit within the year of implementation? Why?
(5) Did implementation generate any funds that had not been anticipated?
(6) Now, after implementation is complete, are there limitations or capabilities of the system which hadn't been expected initially?

These questions should give you every opportunity to anticipate and minimize each problem that is listed below.

Problem 1: The Conversion of Accounts Bottleneck

The conversion of accounts bottleneck is easy to understand. Somebody is going to have to type some information into the computer about every patient in your practice. If you have 3,000 active patients, and if typing the data takes from 5 to 10 minutes per patient, then it will require 250–500 hours of *extra* effort to get the system operating. Sometimes it is difficult to get data assembled and entered in 10 minutes. This time commitment will put tremendous stress on your staff if it is *added* to their current responsibilities. You should hire sufficient temporary help or require sufficient help from the vendor to get the job completely done within 30 days.

Problem 2: The Key Employee Problem

A manual system for billing and management of accounts receivable is the most sophisticated administrative activity in your office. Behind this sophistication there is a key (if not *the* key) employee who understands the system and remembers the exceptions and numerous special situations.

Converting to a computer will make billing and (to some degree) management of the accounts receivable one of the most mundane tasks in your office. You can reasonably expect the computer to cut your accounts receivable by 50–60%. Much of this improvement will come because the computer will decrease exceptions and special situations.

The bottom line is this: your computer system itself probably will undercut the special position of a key employee in your office. This fact will emerge during implementation as a major negative incentive. Your key employee will have every reason to hope consciously or unconsciously that the system will fail.

This antipathy is rarely recognized by the employee, but is manifested by an intolerance for the system's "rigidity," by a tendency for data to be entered wrong and the system blamed. It may be manifested by spells of depression or absenteeism which can affect office morale. It may be manifested by unfortunate accidents or rough handling of equipment.

The problem of the key employee can be anticipated by soliciting the interest of this employee in selecting, implementing, or running the system. Alternatively, you can anticipate that some employees will have problems adapting to an automated office and make plans for phasing them out early in the process.

Problem 3: Inaccurate Data

A variety of pressures will encourage the entry of inaccurate information into the data system as it is implemented. Your employees may suddenly find themselves expected to perform all of their routine chores *and* enter several thousand patient records into the computer (Problem 1). An employee may be unhappy with the new system (Problem 2). An employee who is not a skilled typist may suddenly be asked to type. A piece of equipment could be faulty.

For any of the above reasons, data may be entered incorrectly. If this is common—if patient names are spelled incorrectly, if addresses are wrong, or if unedited data fields are ignored—you will have nothing but problems.

The problem is easy to handle. Beginning on the *first day* print out 1 or 2% of the data that has been entered and check it manually for systematic or careless errors. Alternatively, you may wish to send a printout of each patient's data to that patient for review and correction. In this manner, systematic errors of data entry can be caught early and the source corrected.

Problem 4: System Downtime

In your mind's eye you most probably envision your new computer operating at the time your office is open. In many cases, it will not be operating. For an in-house system, it is not unusual for the system to be "down" (inoperable) 1 or 2 weeks per year. Some smaller systems must be returned to the manufacturer for repair.

Downtime is a routine part of living with computers and is no problem if there are adequate backup procedures in place. Variants include:

(1) Not putting data in the machine which must be used on the day of the visit.

(2) Printing out important data for the next visit within a day or two after the patient leaves the office, for example, keeping it in the patient's chart.

(3) Buying two machines. This practice is increasingly common with microprocessors, but does not work if the power to your office fails. The backup machine can be used for a noncritical function such as word processing during periods of downtime.

Problem 5: System Sclerosis

Chapter 2 deals with evaluating procedures to avoid purchasing a system which is too small or inflexible. Once implemented, however, the system which you choose may simply be inadequate to handle your longer-term requirements. If this happens to you, these are a variety of alternatives available to you.

(1) Use the computer to support manual procedures. Many jobs which you might wish to computerize can also be readily done with *periodic* computer support. For example, a book which cross-indexes patient name, address, and telephone number can be printed out periodically. Data about specific visits can be stored off-line and recalled once or twice a month.

(2) Explore the possibility of expanding your hardware (equipment) without altering software (programming). Many vendors are willing to exchange your equipment for that with higher performance at a nominal cost (perhaps $500 plus the difference in equipment value).

(3) Consider using a microcomputer as a terminal for the larger system, using the microprocessor itself as a word processor, electronic file cabinet, or to monitor your finances. Typically, these functions can be performed using cheaper "canned" programs from the shelf of the computer store. When you phase in a new main computer in 5–7 years, you may choose to just move your micro to the new machine, or leave it connected to both machines for a period to ease the transition.

No attempt has been made here to enumerate all problems which you may encounter. Your best sense of these will be obtained by visiting an operating site prior to your own installation. The problems outlined here *will* occur during implementation in any facility, but can be easily anticipated and minimized with a little forethought.

Chapter **15**

Office Computing and the Right to Privacy

Elmer Gabrieli

Objectives

Discuss relative data security, various security risks, and measures to prevent need to anticipate networking needs in the future. Review measures to keep data anonymous, problems for error correction, and access to data.

Introduction

In the "good old days" (which were perhaps not always that good), the patient did not hesitate to reveal to his doctor even the most personal and confidential information, such as drug-taking habits, drinking, sex life, or intimate family problems. The family physician was the only person involved in the sharing of such sensitive information, and the patient was convinced that the privileged communication would remain confidential.

Medical privacy, in my interpretation, includes the patient, the care-providing physician, as well as the institution (hospital or clinic). All these have their rights. We know that if we were to allow potentially sensitive medical information to go unprotected, it would have a serious negative effect upon our entire health care system.

Modern medicine calls for a health care team, and this team is a growing concern for those who wish to protect the patient's right to privacy.

Computerization of medical data offers immense benefits. It is, however, imperative to install "safe" systems which protect medical privacy.

A frequently encountered misunderstanding in medical computing is the interpretation of the term "data security." It is interpreted by the uninitiated as an absolute guarantee. That is, of course, impossible. No data security system can offer absolute protection. However, we very much want to protect our data to a certain degree, and therefore it seems important to focus on the concept of *relative* data security. I believe clinical medicine will be satisfied if the computerized sensitive data were as "safe" as the (often illegible) notes of the physician kept under lock and key in his private office. If a data security system makes unauthorized access to sensitive data very *difficult*, and if any deliberate violation of medical privacy were adequately *penalized* by statutory measures, we could call computerization relatively "safe."

What Is to Be Protected?

Perhaps initially a physician planning to install a dedicated system in his office may feel that data security is not a major concern. The same secretary and nurse will continue to have access to these computerized data, and actually, the traditional paper record presents a greater risk since it is readable by anyone. However, we should anticipate that medical office computers will follow the trends of the data processing industry. In the future, dedicated office computers will share information with other medical offices, gradually networks will evolve, and this evolution will represent an inherent data security risk to the patients, the physician, and all others involved. Therefore in planning an office data system, it seems important to think in a "forward-compatible" way, and anticipate networking in the future. Privacy is threatened when a person involved can be recognized. Thus, we must focus our efforts on keeping the computerized data anonymous.

The simplest rule is to separate the "header" (all the social identifiers) from the clinical data. This is a time-proven method in clinical medicine. Oral presentation of a case starts: "Mr. G. is a 55-year-old white male . . . ." The same separation of the header from the clinical data can be readily accomplished right from the outset. This results in a list of "keys," the directory of headers. This listing may increase the complexity of the data system, but it seems worth the small extra trouble.

Error Correction

Since electronic patient records will gradually replace handwritten notes, a new set of rules for correction of errors is required. An obvious rule is not to tolerate *direct* record correction. Actually, the software should prohibit direct correction. If any entry errors are made, a pointer should be attached indicating that a correction was recorded later. If this rule of "no corrections" is not a part of the system design, any legal proceeding may be embarrassing. It would undermine the credibility of the electronic record.

Access by Courts

When a patient chooses to file a personal injury claim, and the courts wish to examine the electronic records (in the form of a printout), there is no further problem with the privacy rights of the patient. By seeking legal action, the patient has waived his right to privacy.

Implied Consent

When a patient seeks medical care and provides his clinical history, there is an implied consent for record-keeping by the medical office. Computerization is simply one form of record-keeping. It may be, however, advisable to let the patients know formally that computer technology is being used for record-keeping, and that strict data security measures have been instituted.

Data Security Measures

The *hardware* security measures begin with protection of the terminal. Strict intraoffice operating rules are needed. These rules must be established and enforced by the physician in order to limit the use of the input terminal to the person(s) fully trained to use the system. Tampering with the input terminal should not be allowed. The terminal should be activated only when the data entry person is present. The data security system should be sensitive to any unauthorized attempt to use the terminal.

The *software* security measures should be particularly carefully planned, and the physician should devote the necessary time to fully understand the software security system. The vendor's verbal assurance should not be accepted. Instead, a step-by-step demonstration of the data security system should be an important part of the general presentation by the vendor.

1. User Identification

Each member of the office (including the physician(s)) would have a personal password which is to be changed at least monthly. It is essential to keep the passwords secret, and the physician should play a cardinal role in demanding full cooperation by the staff. Casual handling of passwords will erode the entire security system.

2. Patient Headers

The physician should decide which data should be treated as sensitive elements of the header. The list of these sensitive data should certainly include name, address, and telephone number, but it may well also include employment, insurance data, and other socioeconomic data. The composition of the header should reflect the type of practice, such as urban vs. rural. An ophthalmologist may be less concerned about data security than a family physician or a gynecologist. Each sensitive element in the header should be numbered, and clearly defined. The list of sensitive data should then be incorporated into the system software so that automatic data protection can be implemented.

Retrieval

Rules should be formulated to restrict patient data retrieval to authorized persons for justifiable purposes. Some of these are:

(a) Office visit by the patient
(b) Telephone call from the patient or the family of the patient requiring medical advice
(c) Hospitalization
(d) Billing
(e) Audit
(f) Physician's request

The software should maintain "audit track" on all header-linked retrieval, and the periodic data security report (as part of the routine report to the physician) should tabulate all such retrievals. (The retrievals without the header represent a lesser risk, and may not require an audit trail.)

These retrieval rules are quite different for a solo practice, a group practice with several physicians, or a clinic.

Another major security consideration is display on the screen vs. printed output. Retrieval with display during regular office hours is the lowest-level security risk, whereas a printout (mailed to a third party) may represent the highest risk. The type of practice and operating habits should modify the retrieval rules.

4. Patient Access

Most of us still feel that unlimited and uncritical access to medical records by any patient may not be appropriate, but it is increasingly accepted that patients should be allowed to see their own records, either directly (in most cases), or through the interpretation of the patient's own physician. Confidentiality is violated when a patient's medical data are accessed by another person not a member of the health care team without the expressed permission of the patient. Thus, reading his own records is generally not a problem of confidentiality, as long as the patient accesses his own history and diagnostic studies. However, in certain cases, members of the family, co-workers, friends, and others may be interviewed, and such information may be included in the patient's file. Confidentiality in such cases refers to these interviews when confidential data are given by others rather than the patient. Therefore, such records represent a complex data security problem.

5. Government Access

The principle we believe in is that in clinical medicine the patient's rights and benefits should be the primary consideration. We believe medicine is an altruistic humanitarian profession with the sole purpose being to help, prevent, cure, or alleviate. To use medical data for law enforcement purposes is a particularly controversial issue. For example, using the multistate psychiatric registry listing as part of the gun license issuing routine seems to ignore this principle. Of course, all physicians are in favor of law and order, but medicine should not be an arm of the police. The police should use other methods for screening gun license applications.

It is equally controversial that the list of surgeries done in some hospitals is automatically shared with the Internal Revenue Service. Both the patient's rights and the physician's rights are violated.

A special problem, still insufficiently studied, is the prescription of Schedule II drugs. Patients with painful clinical conditions may be placed on the same list as drug abusers. Computerization makes such data readily retrievable, and the risk is misuse of such data.

6. Access to Data After Office Hours

As a general rule, data security risk is particularly great outside regular office hours. Some physicians keep the office computer up and accessible from a remote site night and day. This step enables them to call up their own office computer from home, or elsewhere, to answer emergency questions. This is a particularly high security risk since the computer can be called up by anyone interested in obtaining medical information. Fortunately, there is ample software technique for high-level data security, even in such situations. Thus, such an arrangement can be made rather safe, if the necessary security measures are implemented.

Concluding Remarks

Computerization, like any other technical advance, offers immense benefits, coupled with inherent risks. The two major risks are loss of data and violation of privacy. Both risks can be kept under sufficient control if the physician is sufficiently informed and insists on adequate safety measures. It is the hope of the author of this chapter that some medical organization will recognize the need for standardization, and "safe" systems will be tested and approved by a team of experts. This would remove the current burden of evaluating the data security systems of various vendors from the individual physician.

Chapter 16

Summary

Byron B. Oberst

Objectives

Emphasize need to plan properly, to think properly, to think ahead 3–5 years, to consider security and privacy, what details to expect from vendors.

Introduction

The American Association for Medical Systems and Informatics has enjoyed putting together this book. We have endeavored to create a How-To-Do-It book on the fundamental approaches to the important considerations which are necessary for a successful venture. (Case Study of the Omaha Children's Clinic, P.C., described in Appendix B, relates all of the points made in the book to an actual office operation.)

The value of thinking through your own office needs prior to beginning is the essence of a successful installation. Include thoughts of how you might want things to be in 3–5 years hence. Critically examine your current office practices and the orderliness of your own habits. If your office is in a state of disarray, then a great deal of tidying up is in order before you begin. You must expect your own habits to improve before you can expect your office personnel to change their routines.

Seek competent consultative assistance so that the equipment and software can be tailored to your needs and ideas. In this manner, there is not so much traumatic dislocation and change. The consultant should be capable of researching out the system to fit your own particular needs inasmuch as possible. This assistance will usually save you time and money in the long haul.

Plan "Beyond Billing" so that the additional accounting and administrative benefits are yours to "have and to hold" until better things come along.

Early in your planning and decison processes give considerable consideration to security and privacy. These factors are vital to you and to your patients. Do not shortchange these considerations.

Your computer consultant should help you prepare to talk with vendors and to evaluate the real contents of the various sale pitches. Be discerning. Document the preliminary discussions by a brief summary of your own understanding of the vendor's equipment and proposal; then, ask the vendor to verify your comprehensions of the proposal. This protection will be of great assistance at contract negotiating time.

Obtain a very competent attorney with some knowledge, expertise, and experience in data processing contracts. Insist on clarification of major issues before you complete your negotiations and long before you consider signing any contracts. These points allow you maximum flexibility and preserve your options. Once the contract is signed, your ability to maneuver and shift is markedly limited—something like zero.

Include your key office personnel who will be directly involved in computer usage early in the planning process. Your decisions will be better formulated using their thoughts and suggestions. They, not you, will be doing most of the work with the computer. Keep the whole process one of a great adventure and exploration to your office staff so that their enthusiasm can be more spontaneous and self-sustaining. The change-over irritations and headaches will be less.

Proceed on a defined timetable and in modular increments. It makes the process easier and less dislocating to work habits.

Allow room for further growth and development so that Health Care Delivery aspects can be included as you deem useful and necessary. By focusing on some of the major health care applications in your preliminary planning processes, there is less likely a chance of not having the capabilities available when you finally decide to add further dimensions to your system. Your computer consultant should be able to assist you in these considerations. Nothing is as frustrating and disappointing than to find that a failure to plan properly makes new applications impossible without major changes and high costs.

Good planning—Good hunting—Good implementation.

Appendix A

Computer Hardware and Software

John M. Long

General Computer Hardware Components

Data Entry and Display Equipment

In its most basic state a computer can only handle logical operations on symbols it can recognize. Therefore information has to be converted into such a format, the so-called machine-readable form, before the logical circuitry can process it. Many devices are used to provide for this man–machine interface. The most common method years ago (actually one still used extensively today) were cards with square holes in them (made famous by the terms "do not fold, spindle, or mutilate"). The most common data entry form today is a standard typewriter keyboard which is often attached to a cathode-ray tube (CRT) so that you can monitor the data being entered. The CRT also allows the computer to display information (i.e., show you what is in its files). The punched cards, the keyboard entry device, and the CRT are only three of many entry and retrieval devices. Another common way of receiving data from a computer is called the printer. It is a glorified typing device which can print large amounts of data in a wide variety of formats on continuous forms.

Permanent and Temporary File Storage Devices

Generally speaking, *information* implies comprehension, whereas *data* is a collection of symbols which transfers information into a recorded form. The terms often lose their distinction. In any event, computers consume and produce high volumes of data and some means must be provided to store this data. Permanent file storage devices are so called because they are permanently attached to or are a part of the computer hardware. Temporary file storage devices can be removed. For either type, permanent or temporary, the contents of the files can also be permanent (set in time) or temporary (changing in time). Many esoteric devices for storing data exist. All essentially store binary numbers in some form.

Coded data is often stored on reels of magnetic tape (including cassettes for a small home computer) and on magnetic discs, which are shaped like a phonograph record. These are sometimes stacked to increase their capacity.

While we are talking about files it might be useful to mention a few computer terms for you to keep in mind. The most basic pieces of data are referred to as bits or bytes, which are combined to form characters (corresponding to letters of the alphabet, numbers, and other symbols such as are found on a typewriter keyboard). A string of characters which constitutes a meaningful piece of information (e.g., patient name) is called a record, a set of records is called a file. A large number of files logically connected so that relationships can be determined are called databases. For example, a set of files which include lists of patient names and addresses, charges generated, bills sent, and payments made might be called a patient accounts database. When the necessary logical circuitry and programs are added to produce bills, acknowledge receipts, determine past due accounts and the like, the system might be called the patient accounts system.

Central Processing Unit and Central Memory

This core of a computer (where the manipulation of data is done following the logical processes built into the computer circuitry) is called the central processing unit (CPU). The CPU modifies the states of the bits of information in its registers as called for by the program. It needs to be able to move these bits around very quickly. In order to do this, special storage capacity is permanently built into its circuitry and is called central memory. Central memory is often built out of circuitry whereas other file storage devices are often magnetic. The capacity of the CPU is usually limited by the amount of central memory available for rapid transfer of data. The amount of central memory is often given in kilos (K), thus a memory with 48K bits has 48,000 bits. When reviewing advertising material, note that 48K bits isn't the same as 48K characters.

General Computer Software Components

Over the past 20 years or so, combinations of logical instructions have been built so that a single line of code (an instruction) in a computer program will call a series of other more basic instructions to accomplish the request. These layers of instructions have reached the point today where a nonprogrammer can use an English-like stylized instruction to get a computer to do its wish. For example, you may have used a terminal in the hospital where you can quickly step through a sequence of options and produce a treatment plan for one of your patients.

As computer languages have become more sophisticated and complex, we have placed certain related operations together into categories thus:

Compilers are a special set of "translators" which receive a "source" code set of statements somewhat comprehensible to humans (like FORTRAN, COBOL, etc.) and convert them into the tedious detailed logical manipulators comprehensible to the computer. The "object" code set of statements which result is the program which will actually cause the computer to perform as desired.

Operating systems are a collection of internal computer management functions as well as a large number of commonly done "housekeeping" functions which are kept available at all times so that each program can use them as needed. The operating system does such things as keep track of each program, set up files, keep track of where the files are located, move them around as needed, call in data off of tapes and discs, send data out to printers, activate the printers, and the like.

File handling and file management turns out to be a major task when doing things on a computer. Some approaches are far more efficient than others. These functions are also packaged into sets of efficient routines called file handling and file management software. Often these are included as a subset of the operating system.

Software to allow the computer to receive, interpret, and process data entered from a variety of sources including tapes, discs, and remote terminals comprises another major segment of software. Software to perform a variety of logical checks for errors, so-called "edits," are usually included in the so-called *information storage* software.

Similarly, combinations of routines to allow us to retrieve, format, and display data found in the computer files represents another major software grouping. *Report* generators, which allow the data to be formatted in a meaningful way, are a part of the so-called *information retrieval* software.

Computer Systems Design

Today, the hardware and software are combined to produce automation capacity which can do the task described in this book with almost all of the details mentioned above being transparent to you, the user.

A Case Study of Computer Applications to the Omaha Children's Clinic, P.C., Utilizing Bishop Clarkson Memorial Hospital's Private Practice System

Byron B. Oberst

Historical Background

The author has collaborated with Bishop Clarkson Memorial Hospital for the past thirteen years in adapting the computer to office practice. The administrator of this tertiary care facility, James Canedy, has the philosophy that a hospital exists to provide outstanding care to patients and to supply the needed support facilities to the medical staff to render this care.

The Private Practice System was conceived by the author (see References 1 through 9) and developed by the Clarkson Computer Staff. Plans were developed so that as in-hospital computer applications were developed, these applications were subsequently adapted to the office practice system. The Omaha Children's Clinic, P.C. (OCCPC) was used as a laboratory for testing and refining these various computer applications. The Private Practice System is structured so that the customization of various types of medical practices can be easily accomplished. Among other medical offices this includes a one-man solo practitioner, a two-man family practice, a three-man surgical office, and a six-man pediatric office.

The Private Practice System is structured in a modular, stand-alone fashion so that the physician can utilize a complete package of the various component parts: (1) Accounting Module; (2) Administrative Module; (3) Insurance Module; and (4) Health Care Delivery Module as currently developed depending upon his or her needs. Costs are reasonable and based upon the modules used in the number of account guarantors kept in the on-line active file. The various modules will be briefly described.

Accounting Module

The Accounting Module justifies all of the other computer applications through cycle billing, aging of accounts receivable, defined collection procedures, and other elements. Accounts are purged to off-line tapes if there is no activity within the previous eighteen months. Dollar control and volume of work load has steadily increased as the OCCPC has grown. Cost accounting is utilized to determine physician fees and procedure charges. Excellent fiscal stability and financial planning has been the end result.

Administrative Module

The Administrative Module applies sound management principles utilizing physicians designated as Cost Centers as a means of allocating income generation and assigning fixed and variable costs in a systematic manner. Budget forecasting, practice statistics report generation, and other elements are all part of the Private Practice System. There is limited word processing available including messages on office statements and label generation. Satellite capabilities are easily applied along with individual and collective physician production identifications. The OCCPC's satellite functioned well with a minimum of personnel because of the computer. Due to other circumstances, the satellite was later discontinued.

Insurance Module

Insurance form generation, identification of various insurance companies, utilization of some insurance companies' health maintenance programs, state and federal welfare applications, and the OCCPC office patient yearly budget plans are some of the applications of the computer in the Private Practice System. Use of CPT-4 and ICD-9 are utilized for uniform coding purposes.

Health Care Delivery Module

Computer applications in this area add a major dimension to the capabilities of patient care within the OCCPC.

The health care applications utilizing computer assistance offers to the physicians of the Omaha Children's Clinic, P.C. opportunities to improve health care delivery. The following describes the Omaha Children's Clinic, P.C.'s Medical Synopsis.

The Medical Synopsis of a patient's ongoing medical record is the central core of improved health care delivery, improved quality control and the accumulation of data related to treatment effectiveness. The following items are included in the Synopsis:

1.0 Demographic data
2.0 Physical—anthropometic measurements including height, weight, blood pressure, heart rate, respiration rate, temperature, head circumference, chest circumference, and skin-fold thickness
3.0 Patient and family outcome functional capacity
4.0 Brief family history
5.0 Brief patient history
6.0 Listing of immunizations with dates
7.0 Potential medical problem list
8.0 Adverse reactions due to medications and other materials
9.0 Problem list including active and inactive medical problems with associated comments and designated status of the problem. The status of an active problem is indicated by the following codes:

9.1 Episodic	9.11 Terminal
9.2 Long term	9.12 Worsening
9.3 Chronic	9.13 Initial
9.4 Intermediate	9.14 Potential
9.5 Improving	9.15 Adverse effect
9.6 Inactive	9.16 Removed
9.7 In evaluation	9.17 Repaired
9.8 Moderate	9.18 Case closed
9.9 Serious	9.19 Case reopened
9.10 Complex	

10.0 In-office laboratory listings of results including X-rays and EKG reports
11.0 Out-of-office laboratory listings of results including consultations, X-rays, EKGs, and CAT scan reports
12.0 Medication listings including dosage, intervals, and duration. Listing of the special care treatment programs are the following:

12.1 Low fat diet	12.4 Special diet
12.2 Reduction diet	12.5 Lactose free diet
12.3 Ulcer diet	12.6 Diabetic diet

12.7   Other   12.13 Physical therapy-appliance
12.8   High protein diet   12.14 Gluten free diet
12.9   Elimination diet   12.15 Psychotherapy-counseling
12.10 Rotation diet   12.16 Group therapy
12.11 General diet   12.17 Sinusitis routine
12.12 Environmental control

13.0 A listing of the types of visits and procedures are the following:
   13.1 Initial visit
   13.2 Annual health review with age bracket of child or youth
   13.3 Consultation with annual health review with age bracket of child
        or youth and with added time modules for associated major
        medical problems
   13.4 Consultations with time modules
   13.5 Medical or family conference with time modules
   13.6 Associated procedures performed

All problems, tests, procedures, and office visits are tied together using the same related problem number. This linkage provides the capability for auditing the quality of care, the determination of treatment compliance, and treatment cost effectiveness. This latter capability is not available to the OCCPC as yet, but all plans and parameters have been developed and are awaiting a priority assignment within the Clarkson Hospital overall computer development plan which has been delayed due to equipment change-over.

The Medical Synopsis provides a personalized patient health index system including a hard copy for the office and one for the patient's family and identifies the appropriate disease flow sheets needed. Other capabilities are the following:

1.0 Indicates patient recall needs for such disease states as hypertension, diabetes, attention deficit disorder, and others
2.0 Indicates types of medical care given
3.0 Includes office satellite capabilities
4.0 Gives opportunities for surveillance on such elements as the completion of immunizations, the need for annual health review or other types of visits, and the utilization of the patient-family education programs
5.0 Insures the collection and retrieval of data and statistics
6.0 Allows for the transmittal of vital data to other approved users
7.0 Arranges case disposition including:
   7.1 Logging of parent-patient education with "Guided Growth" advice, dispensing of associated printed materials, utilization of audiovisual material, posting parent-teen lectures, and other education or health supervision data
   7.2 Identifying the needed return visits or phone calls for progress and/ or return to the referring M.D.
   7.3 Providing for a listing of associated current references and articles as applied to a specific problem disease entity

The OCCPC has previously utilized an Optical Scanning Device (OPSCAN) method to develop a newborn hospital dismissal summary but the machine was not utilized enough to justify the costs. For future use, the OCCPC is in the process of developing detailed automated histories for various age periods which identify potential and/or active problem areas and indicate the preventive or corrective medical action plan which can be instituted. The histories for family, infancy, and preschool periods are currently manually operational. In the future, the following capabilities will be added:

1.0 A patient recall and surveillance system
2.0 A modular wave appointment program
3.0 Data recall
4.0 Quality control and audits of patient care

The computer has given to the Omaha Children's Clinic, P.C. the management control and capabilities for realistic long range planning based upon firm data and excellent monetary management.

Advantages

1.0 Simplified accural of large amounts of useful data
2.0 Delineation and summarization of pertinent facts regarding a patient's medical history and background
3.0 Ease of identification of the patient's medical problem areas
4.0 Improved follow-up recall and surveillance of medical care
5.0 Potential capabilities for clinical investigation
6.0 Potential for assistance in the relicensure process if necessary
7.0 Potential for office auditing and quality control of patient care
8.0 Potential for determining cost effectiveness of medical care and treatment
9.0 Capabilities of the problem-oriented medical record approach to patient care
10.0 Protection of higher level sensitive patient data
11.0 Ease of transmission of pertinent data to the other authorized health care units
12.0 Frequently updated hard copy medical synopsis for the patient's own files

Disadvantages

1.0 Requires change in habits and the way of doing things by the physician and personnel
2.0 Changeover costs
3.0 Problem of input errors and subsequent need for correction
4.0 Headaches, irritations, and confusion during the conversion period
5.0 Problems of confidentiality

Needs

1.0 A structured and formatted approach to health care practices
2.0 Useful and appropriately designed software including charge slips and other data input forms
3.0 Patience and persistence

Conclusion

The Bishop Clarkson Memorial Hospital Private Practice System has helped the Omaha Children's Clinic, P.C. develop into an ultramodern pediatric and adolescent medical health care delivery system. It has provided the foundation for future planning, growth, and innovations. We have just completed developing our second Ten-Year Long-Range Plan based upon accumulated sound data. This Long-Range Plan will be instituted in small predetermined increments through an Annual Operating Plan as deemed appropriate. The future appears very exciting and satisfying.

References

1. Oberst, B.B.: *Practical Guidance for Office Pediatric and Adolescent Practice*, Springfield, Charles C Thomas, 1973.
2. Oberst, B.B.: The Development of Norms and Guidelines for Office and Hospital Medical Care; *Society for Computer Medicine*; Proceedings of the Fourth Annual Conference of the Society for Computer Medicine, 1974.
3. Oberst, B.B.: A Private Practice System Utilizing the Computer and the Problem Oriented Record, *Computer Medicine*, October 1975.
4. Oberst, B.B.: A Total Health Care System As Viewed by A Private Practitioner, I: A Composite Overview, *Pediatrician* 4:176–184, 1975.
5. Oberst, B.B.: A Total Health Care System As Viewed by A Private Practitioner, II: A Conceptual Design, *Pediatrician* 4:372–383, 1975.
6. Oberst, B.B.: A Total Health Care System As Viewed by a Private Practitioner, III: A Progress Report, *Pediatrician* 4:383–392, 1975.
7. Oberst, B.B.: The Need for a Commonality of Purpose, Design, and Application of Computers Within the Medical Specialty Organization, *Journal of Clinical Computing*, Vol. 5, No. 2, 1976, pp. 139–140.
8. Oberst, B.B.: Application of the Computer to Office Practice, *Society of Computer Medicine*, Proceedings of the Eighth Annual Conference of the Society for Computer Medicine, 1978.
9. Oberst, B.B.: The Administrative Anatomy of the Pediatric Practice of the Future, *Clinical Pediatrics*, Vol 18, No. 1, January 1979, pp. 9–11.

Index